Security and Privacy in Wireless and Mobile Networks

Special Issue Editors

Georgios Kambourakis
Félix Gómez Mármol
Guojun Wang

MDPI • Basel • Beijing • Wuhan • Barcelona • Belgrade

MDPI

Special Issue Editors

Georgios Kambourakis
University of the Aegean
Greece

Félix Gómez Mármol
University of Murcia
Spain

Guojun Wang
Guangzhou University
China

Editorial Office
MDPI AG
St. Alban-Anlage 66
Basel, Switzerland

This edition is a reprint of the Special Issue published online in the open access journal *Future Internet* (ISSN 1999-5903) from 2017–2018 (available at: http://www.mdpi.com/journal/futureinternet/special_issues/Wireless_Mobile_Networks).

For citation purposes, cite each article independently as indicated on the article page online and as indicated below:

Lastname, F.M.; Lastname, F.M. Article title. *Journal Name* **Year**, *Article number*, page range.

First Edition 2018

ISBN 978-3-03842-779-7 (Pbk)
ISBN 978-3-03842-780-3 (PDF)

Table of Contents

About the Special Issue Editors

Georgios Kambourakis received a Diploma in Applied Informatics from the Athens University of Economics and Business (AUEB) and a Ph.D. in Information and Communication Systems Engineering from the Department of Information and Communications Systems Engineering of the University of the Aegean. Currently, Dr. Kambourakis is an Associate Professor at the Department of Information and Communication Systems Engineering, University of the Aegean, Greece, and the director of Info-Sec-Lab. His research interests are in the fields of mobile and wireless networks security and privacy, VoIP security, IoT security and privacy, DNS security, and security education, and he has more than 120 refereed publications in the above areas. More info at: www.icsd.aegean.gr/gkamb.

Félix Gómez Mármol is a senior researcher in the Department of Information and Communications Engineering at the University of Murcia, Spain. His research interests include cybersecurity, the Internet-of-Things, machine learning and bio-inspired algorithms. He received a M.Sc. (Honors) and Ph.D. (Honors) in computer engineering from the University of Murcia. He has published 30 articles in journals indexed in the JCR, 13 at international conferences and two book chapters, accruing a total of over 1280 citations (h-Index 16). He accrued five international patents exploited by the NEC Corporation, as well as two open source projects, with over 9200 and 1400 downloads, respectively. He has participated as a Technical Program Committee member at more than 75 international conferences, also serving 15 times as general co-chair, program co-chair, publicity co-chair or industry co-chair. Additionally, he has collaborated as an editorial board member for six international journals and 11 times as a guest editor for special issues in international journals. He also contributed to three national research and development contracts, five national R&D projects and 11European research projects, acting as main investigator from NEC, as well as work package leader, for three of them. More info at: http://webs.um.es/felixgm

Guojun Wang received a B.Sc. degree in Geophysics, a M.Sc. degree in Computer Science, and a Ph.D. degree in Computer Science from the Central South University, China (CSU), in 1992, 1996, 2002, respectively. He is a Pearl River Scholarship Distinguished Professor of Higher Education in Guangdong Province, a Doctoral Supervisor at the School of Computer Science and Educational Software, Guangzhou University, China (GU). He had been a Professor at Central South University, China; an Adjunct Professor at Temple University, USA; a Visiting Scholar at Florida Atlantic University, USA; a Visiting Researcher at the University of Aizu, Japan; and a Research Fellow at the Hong Kong Polytechnic University, Hong Kong. His research interests include artificial intelligence, big data, cloud computing, mobile computing, trustworthy/dependable computing, cyberspace security, recommendation systems and mobile healthcare systems. He has published more than 300 technical papers and books/chapters in the above areas. His research is supported by the Key Project of the National Natural Science Foundation of China, the National High-Tech Research and Development Plan of China (863 Plan), the Ministry of Education Fund for Doctoral Disciplines in Higher Education, the Guangdong Provincial Natural Science Foundation, the Hunan Provincial Natural Science Foundation, the Hunan Provincial Science and Technology Program, and the Changsha Science and Technology Program. His research is also supported by talent programs including the Program for Pearl River Scholarship Distinguished Professor of Higher Education in Guangdong Province, the Program for New Century Excellent Talents in University, and the Hunan Provincial Natural Science Foundation of China for Distinguished Young Scholars. He is an associate editor or on editorial board of international journals including IEEE Transactions on Parallel and Distributed Systems (TPDS), Security and Communication Networks (SCN), International Journal of Parallel, Emergent and Distributed Systems (IJPEDS), and International Journal of Computational Science and Engineering (IJCSE). He has served as guest editor-in-chief or guest co-editor for international journals including IEEE Transactions on Parallel and Distributed Systems (TPDS), Journal of Computer and System Sciences (JCSS), and IEICE Transactions on Information and Systems. He has served as general co-chair or program co-chair for a number of international conferences including IEEE Smart World Congress 2018, ISPA 2017, IUCC 2017, ISPEC 2016, APSCC 2016, PRDC 2015, ICA3PP 2015,

HPCC 2013, CSS 2013, MobiQuitous 2013, ICA3PP 2012, and ATC 2009. He was the leading steering chair of the IEEE International Conference on Trust, Security and Privacy in Computing and Communications (TrustCom 2011), and the leading steering chair of the International Conference on Security, Privacy and Anonymity in Computation, Communication and Storage (SpaCCS 2016). He is a member of IEEE (2010-), a member of ACM (2011-), a member of IEICE (2011-), a distinguished member of CCF (2013-), Vice Chairman of the Intelligence Engineering Society of Guangzhou (2017.9-), and an executive member of the council of Hunan Provincial Association of Computers (2011–2016).

future internet

MDPI

Editorial

Security and Privacy in Wireless and Mobile Networks

Georgios Kambourakis [1,*], **Felix Gomez Marmol** [2] and **Guojun Wang** [3]

[1] Department of Information and Communication Systems Engineering, University of the Aegean,
    83100 Karlovasi, Samos, Greece
[2] Department of Information and Communications Engineering, University of Murcia, 30100 Murcia, Spain;
    felixgm@um.es
[3] School of Computer Science and Educational Software, Guangzhou University, Guangzhou 510006, China;
    csgjwang@gzhu.edu.cn
[*] Correspondence: gkamb@aegean.gr; Tel.: +30-227-308-2256

Received: 6 February 2018; Accepted: 7 February 2018; Published: 9 February 2018

Currently, at the dawn of 5G networks, and the era of the Internet-of-Things, wireless and mobile networking is becoming increasingly ubiquitous. In this landscape, security and privacy turn into decisive factors. That is, the mobile and wireless ecosystem is an ideal playground for many perpetrators: (a) handheld devices are used for critical tasks, such as e-commerce, bank transactions, payments, application purchases, as well as social interaction; (b) such devices uniquely identify their users and store sensitive and detailed information about them; and (c) despite all their sophistication, native security mechanisms of mobile operating systems can be bypassed, and several wireless interfaces and protocols have been proven to be vulnerable to attack. As the attacker is given so many alternative entry points for penetration, the creation of assaults against the end-user and the underlying systems and services have been augmented, both in amount, as well as in matters of complexity. It is, therefore, imperative that new and advanced security and privacy-preserving measures are deployed.

To cope with the aforementioned challenges, this special issue has been dedicated to the security and privacy aspects of mobile networks, wireless communications, and their apps. Particularly, apart from network and link layer security, the focus is on the security and privacy of mobile software platforms and the increasingly differing spectrum of mobile or wireless apps. Via both invited and open call submissions, a total of nineteen papers were submitted and nine have been finally accepted. Each manuscript underwent a rigorous review process involving a minimum of three reviews. All the accepted articles constitute original research work addressing a variety of topics pertaining to the above-mentioned challenges.

The first article by Wenjuan Li, Weizhi Meng and Lam For Kwok [1], focuses on collaborative intrusion detection networks (CIDN), which allow intrusion detection system nodes to exchange data with each other. The authors deal with insider attacks which typically are more difficult to identify. Particularly, by examining challenge-based CIDNs, they analyze the influence of advanced on-off attacks, where the attacker responds truthfully to one IDS node but behaves maliciously to another. The authors report results from two experiments using both simulated and real CIDN environments.

The work by Rezvan Almas Shehni, Karim Faez, Farshad Eshghi and Manoochehr Kelarestaghi [2], copes with Sybil types of attacks in mobile Wireless Sensor Networks (WSN), and proposes a computationally lightweight watchdog-based algorithm for detecting it. According to the authors' algorithm, the watchdog nodes collect detection information, which is then passed to a designated node for processing and identifying Sybil nodes. The highlights of their algorithm are the low communication overhead, and a fair balance between true and false detection rates. These qualities are proved via simulation and comparison against recent watchdog-based Sybil detection algorithms.

End-user privacy protection in smart home applications is the topic of the article contributed by Jingsha He, Qi Xiao, Peng He, and Muhammad Salman Pathan [3]. Given that attacks do not necessarily need access to the cipher, but can be mounted by simply analyzing the frequency of radio signals or the timestamp series, the authors argue that legacy encryption methods cannot satisfy the needs of privacy protection in such applications. Therefore, the daily activities of the people living in a smart home are at stake. To obfuscate the patterns of daily routines of smart home residents, they propose an adaptive method based on sample data analysis and supervised learning, which allows them to cope with fingerprint and timing-based snooping types of attacks. Via experimentation, the authors demonstrate that their method supersedes similar proposals in terms of energy consumption, latency, adaptability, and degree of privacy protection.

Radio Frequency Identification (RFID) systems are inherently prone to attacks because of the wireless nature of the communication channel between the reader and a tag. To protect the privacy of tags, the work by Zhibin Zhou, Pin Liu, Qin Liu and Guojun Wang [4] investigates ways of ensuring the tag's information security and providing guarantees that the system generates reliable grouping-proof. The authors note that since the verification of grouping-proof is typically done by the verifier, the reader is able to submit bogus proof data in the event of Deny of Proof attack. To remedy this issue, they propose an ECC -based, off-line anonymous grouping-proof protocol, which authorizes the reader to examine the validity of grouping-proof without being aware of the identities of tags. The protocol is examined in terms of both security and performance, showing that it can resist impersonation and replay attacks against the tags.

In the mobile app ecosystem, Pierpaolo Loreti, Lorenzo Bracciale and Alberto Caponi [5] stress that push notifications may lead to loss of end-user privacy. For instance, social networking apps use such notifications extensively (e.g., friendship request, tagging, etc.) via real-time channels. However, even in cases where the confidentiality of the channel is preserved, action anonymity may fail. That is because the actions that trigger a notification and the reception of the corresponding message can be uniquely correlated. They pinpoint that even when pseudonyms are in play, this situation can be exploited by attackers to reveal the real identity of the user of a mobile device. The authors call this situation a "push notification attack", and demonstrate that it can be exercised in an online or offline fashion.

The work by Stylianos S. Mamais and George Theodorakopoulos [6] deals with Online Behavioural Advertising (OBA). Concentrating on security, privacy, targeting effectiveness, and practicality, they categorize the available ad-distribution methods and identify their shortcomings. Based on opportunistic networking, they also propose a novel system for distributing targeted adverts in a social network. The highlights of this system are that it does not require trust among the users, and it is low in memory and bandwidth overhead. Moreover, their system blocks evil-doers from launching impersonation attacks and altering the ads with the intention of spreading malicious content. The same authors in [7] note that ad-Networks and publishers service commissions can be forged by non-human actors via the injection of fictitious traffic on digital platforms. This situation leads to financial fraud. Using opportunistic networking and a blockchain technology, they proposed an advert reporting system which is capable of identifying authentic Ad-Reports, i.e., those created by honest users. This is decided by examining, in a privacy-preserving way, the user's patterns when accessing adverts on their mobile device.

The security risks due to design shortcomings and vulnerabilities related to end-user behavior when interacting with mobile devices is the focus of the work by Vasileios Gkioulos, Gaute Wangen and Sokratis K. Katsikas [8]. They present the results of a survey conducted across a multinational sample of security professionals and compare them against those derived from their earlier study over the security awareness of digital natives (young people, born in the digital era). This has been done in an effort to identify differences between the conceptual user-models that security experts utilize in their professional tasks and user behavior. The main result is that, while influences from personal perceptions and randomness are not insignificant, the experts' understanding of the user behaviour does not follow a firm user-model.

The article by Andrea Guazzini, Ayca Sarac, Camillo Donati, Annalisa Nardi, Daniele Vilone and Patrizia Meringolo [9] it built around a very interesting observation: the ICT revolution changes our world and is having a crucial role as a mediating factor for social movements and political decisions. Moreover, the perception of this new environment (social engagement, privacy perception, sense of belonging to a community) may differ even in similar cultures. Motivated by the changes that have occurred due to the introduction of the web, the authors explore via a questionnaire instrument the inter-relations between the constructs of sense of community, participation and privacy compared with culture and gender. Their study took into account 180 participants from Turkey and Italy, with the aim to highlight the cultural differences in the perception of the aforementioned constructs. The analysis of results takes into consideration the recent history of both countries in terms of the adoption of new technologies, political actions, and protest movements.

Author Contributions: All authors contributed equally to this editorial.

Conflicts of Interest: The authors declare no conflict of interest.

References

1. Li, W.; Meng, W.; Kwok, L.F. Investigating the Influence of Special On-Off Attacks on Challenge-Based Collaborative Intrusion Detection Networks. *Futur. Internet* **2018**, *10*, 6, doi:10.3390/fi10010006.
2. Almas Shehni, R.; Faez, K.; Eshghi, F.; Kelarestaghi, M. A New Lightweight Watchdog-Based Algorithm for Detecting Sybil Nodes in Mobile WSNs. *Futur. Internet* **2018**, *10*, 1, doi:10.3390/fi10010001.
3. He, J.; Xiao, Q.; He, P.; Pathan, M.S. An Adaptive Privacy Protection Method for Smart Home Environments Using Supervised Learning. *Futur. Internet* **2017**, *9*, 7, doi:10.3390/fi9010007.
4. Zhou, Z.; Liu, P.; Liu, Q.; Wang, G. An Anonymous Offline RFID Grouping-Proof Protocol. *Futur. Internet* **2018**, *10*, 2, doi:10.3390/fi10010002.
5. Loreti, P.; Bracciale, L.; Caponi, A. Push Attack: Binding Virtual and Real Identities Using Mobile Push Notifications. *Futur. Internet* **2018**, *10*, 13, doi:10.3390/fi10020013.
6. Mamais, S.S.; Theodorakopoulos, G. Private and Secure Distribution of Targeted Advertisements to Mobile Phones. *Futur. Internet* **2017**, *9*, 16, doi:10.3390/fi9020016.
7. Mamais, S.S.; Theodorakopoulos, G. Behavioural Verification: Preventing Report Fraud in Decentralized Advert Distribution Systems. *Futur. Internet* **2017**, *9*, 88, doi:10.3390/fi9040088.
8. Gkioulos, V.; Wangen, G.; Katsikas, S.K. User Modelling Validation over the Security Awareness of Digital Natives. *Futur. Internet* **2017**, *9*, 32, doi:10.3390/fi9030032.
9. Guazzini, A.; Sarac, A.; Donati, C.; Nardi, A.; Vilone, D.; Meringolo, P. Participation and Privacy Perception in Virtual Environments: The Role of Sense of Community, Culture and Gender between Italian and Turkish. *Futur. Internet* **2017**, *9*, 11, doi:10.3390/fi9020011.

future internet

MDPI

Article

Investigating the Influence of Special On–Off Attacks on Challenge-Based Collaborative Intrusion Detection Networks [†]

Wenjuan Li [1], Weizhi Meng [2],*and Lam For Kwok [1]

[1] Department of Computer Science, City University of Hong Kong, Hong Kong, China;
    wenjuan.li@my.cityu.edu.hk (W.L.); cslfkwok@cityu.edu.hk (L.F.K.)
[2] Department of Applied Mathematics and Computer Science, Technical University of Denmark,
    2800 Kongens Lyngby, Denmark
* Correspondence: weme@dtu.dk; Tel.: +45-4525-3068
† A preliminary version of this paper has been presented at the 12th International Conference on Green,
    Pervasive, and Cloud Computing (GPC), 2017; pp. 402–415.

Received: 15 December 2017; Accepted: 3 January 2018; Published: 8 January 2018

Abstract: Intrusions are becoming more complicated with the recent development of adversarial techniques. To boost the detection accuracy of a separate intrusion detector, the collaborative intrusion detection network (CIDN) has thus been developed by allowing intrusion detection system (IDS) nodes to exchange data with each other. Insider attacks are a great threat for such types of collaborative networks, where an attacker has the authorized access within the network. In literature, a challenge-based trust mechanism is effective at identifying malicious nodes by sending challenges. However, such mechanisms are heavily dependent on two assumptions, which would cause CIDNs to be vulnerable to advanced insider attacks in practice. In this work, we investigate the influence of advanced on–off attacks on challenge-based CIDNs, which can respond truthfully to one IDS node but behave maliciously to another IDS node. To evaluate the attack performance, we have conducted two experiments under a simulated and a real CIDN environment. The obtained results demonstrate that our designed attack is able to compromise the robustness of challenge-based CIDNs in practice; that is, some malicious nodes can behave untruthfully without a timely detection.

Keywords: intrusion detection; collaborative network; on–off attack; challenge-based mechanism; trust computation and management

---

1. Introduction

The major goal of an intrusion detection system (IDS) is to identify any signs of suspicious activities in either systems or networks [1]. IDSs are widely adopted in various organizations and can be generally classified into two groups: *host-based IDSs (HIDSs)* and *network-based IDSs (NIDSs)*. The HIDS identifies malicious events for an end system or application by monitoring local events and states. The NIDS focuses on network environments and detects potential attacks by monitoring and examining traffic outside the demilitarized zone (DMZ) or within an internal network [2]. Further, there are usually two major detection methods for a typical IDS, namely, the signature-based detection approach and the anomaly-based detection approach.

A signature-based IDS detects suspicious events by comparing incoming payloads with stored signatures (called rules), while an anomaly-based IDS detects malicious events through identifying significant deviations between the current behavioral profile and the normal behavioral profile. A normal profile is used to describe the characteristics of applications and connections via monitoring for a period of time [1]. With the increasing complexity of current intrusions, it is found that a single or isolated IDS would not work effectively in a complicated scenario [3,4]. These attacks may cause great damage if they cannot be detected timely; that is, they may cause the entire network to be paralyzed. With the purpose of improving the detection accuracy of single IDSs, research has been made for collaborative intrusion detection networks (CIDNs), which enable different IDS nodes to collect and exchange data with each other [4]. The collaborative nature of CIDNs can help to optimize the capability of an IDS; however, insider attacks are one great threat that can significantly degrade the security level of the whole network [3]. As a result, there is a need to implement additional mechanisms to protect a collaborative environment itself.

Building appropriate trust-based mechanisms is a promising solution to protect CIDNs against insider threats. For this purpose, Fung et al. [5] developed a kind of challenge-based trust mechanism (or *challenge mechanism*) for CIDNs, which utilizes *challenges* to evaluate the reputation of IDS nodes. A challenge may contain some predefined alarms requesting the target node to rank the severity. Because the testing node generates the challenge (i.e., extracting from its database), it knows the alarm severity in advance. The reputation of an IDS node can be judged according to the satisfaction level between the expected answer and the received feedback. A line of relevant studies (e.g., [5–7]) have proven that the challenge mechanism can be robust against common insider attacks, like collusion attacks, in which several adversarial nodes work together to provide fake alarm information to target nodes, aiming to degrade the detection effectiveness.

Motivations. The challenge mechanism has shown good performance against common insider attacks, but it depends heavily on two major assumptions: (1) it is hard for an IDS node to distinguish between a challenge and a normal message; (2) malicious nodes would always send untruthful feedback. In a practical implementation, these two assumptions are not realistic in most cases, as adversarial nodes can behave in a much more dynamic and complicated way [8,9]. As a result, because of these assumptions, challenge mechanisms may become problematic under some advanced attacks. As an example, Li et al. [8] designed an advanced attack, named the *passive message fingerprint attack* (PMFA), which could help to distinguish between a challenge and normal messages. Under the PMFA, an IDS node can send untruthful answers to normal requests without decreasing their trust values.

Contributions. In this work, our motivation is to investigate the influence of a special on–off attack (SOOA), which is able to behave normally to one node while sending untruthful answers to another node. Differently from the previous version [10], this work further evaluates the attack performance of the SOOA in a real network environment. The contributions of this work are listed below:

- We first describe the high-level architecture of a typical challenge-based CIDN with the adopted assumptions and then investigate the influence of the SOOA, which can behave normally to one IDS node while responding maliciously to another node. In this case, trust computation in the third node may be affected, as it may receive the opposite judgement from its partner nodes.
- To investigate the attack performance, we have performed two experiments under a simulated and a real CIDN environment. Our results demonstrate that the SOOA has the potential to greatly affect the trust computation of IDS nodes; that is, some malicious nodes can keep their reputation without timely detection. Finally, we discuss some countermeasures and solutions.

Different from the previous work [10], this work both further evaluates attack scenarios and has performed an evaluation in a real CIDN environment. We acknowledge that challenge mechanisms are a promising solution to safeguard CIDNs against malicious insider nodes. The purpose of our work is to attract more research efforts to enhance the application of challenge mechanisms in practical scenarios.

The remaining parts are organized as follows. Section 2 presents a set of related work regarding trust management in distributed IDS networks. Section 3 introduces the architecture of challenge-based CIDNs and analyzes the adopted assumptions. Section 4 describes how (SOOA) works in a challenge-based CIDN and discusses two scenarios as a study. Section 5 describes two major experiments under a simulated and a real CIDN environment. Finally, Section 6 concludes our work with future directions.

2. Related Work

Collaborative intrusion detection systems/networks are developed to boost the accuracy of a separate detector, which usually has less information about the protected environment. This collaborative network enables various IDS nodes to request and collect data from other nodes. However, the collaborative nature renders it vulnerable to insider attacks, in which intruders are inside the network. To protect distributed systems and collaborative networks against malicious nodes, establishing a proper trust-based intrusion detection mechanism is desirable.

Trust-Aware Mechanism

Trust management has been widely studied in literature. Duma et al. [3] described a P2P-based overlay IDS, which utilizes a trust engine to handle alarms and an adaptive scheme to calculate reputation. More specifically, the former is used to filter out alerts sent by untrusted or low-reputation nodes, while the latter can calculate the reputation of nodes by considering their past behaviors. Meng et al. [11] recently proposed a Bayesian inference-based trust mechanism to identify untruthful nodes for medical smartphone networks. The evaluation showed that their approach could quickly identify malicious nodes in real scenarios. For some other related works, we refer to [12–18].

Challenge-Based Trust Mechanism

How to design an appropriate trust management in CIDNs remains an issue. For this purpose, Fung et al. [5] designed a challenge-based trust mechanism, which sends challenges to evaluate the reputation of an IDS node. The trustworthiness of a node can be derived according to the received answers. At first, they described a detection framework based on HIDSs, in which each HIDS node could judge the trustworthiness of others on the basis of the difference between the sent challenges and the received answers. They further utilized a forgetting factor to emphasize the recent feedback [6]. Then, they enhanced their mechanism with a Dirichlet-based model, which allows for the evaluation of the reputation of IDS nodes by considering their mutual behavioral events [7]. In the evaluation, they mainly evaluated their model for challenge-based CIDNs in some simulated environments. The mechanism was found to have strong scalability properties and to be robust against common insider threats.

Advanced Insider Attack

Current intrusions have become more complex, and many research studies have moved to advanced attacks. Li et al. [8,19] developed an advanced collusion attack, named *passive message fingerprint attack* (PMFA), which allows several malicious nodes to exchange received data and distinguish normal requests passively. Experimental results indicated that the PMFA enabled IDS nodes to give untruthful answers to normal requests without decreasing their trust values. Similarly, Meng et al. [9] also developed an advanced collusion attack, called the *random poisoning attack*, which enables a node to provide malicious answers with a predefined possibility. They performed two experiments under both simulated and real environments, and it was found that this attack could compromise the robustness of challenge-based CIDNs.

Mechanism Improvement

To enhance the mechanism performance, Li et al. [20] pointed out that distinct IDS nodes may not have the same detection capabilities. Some nodes could have a higher or lower level of sensitivity for the detection of some particular intrusions. As an example, the number of signatures can decide whether an IDS node has a stronger capability of identifying a kind of virus. That is, a node can be more accurate in identifying such a threat if it has a larger set of relevant signatures. On the basis of this observation, they proposed *intrusion sensitivity* (IS), which could be used to measure the detection sensitivity of an IDS node in terms of particular intrusions. They further proposed a trust management approach by means of IS, through automating the allocation of IS with machine learning techniques in real-world applications [21,22]. Pollution attacks are a kind of insider threat that allow a set of malicious nodes to work collaboratively to give fake alarm information to the target node. Li and Meng [23] conducted a study to explore the influence of IS on the detection of pollution attacks. It was found that this notion can help to detect malicious nodes quickly by emphasizing the impact of expert nodes.

3. Challenge-Based CIDNs

3.1. Background

To protect collaborative networks against insider attacks, many trust-based approaches have been proposed [24]. Challenge mechanisms are one effective approach to point out unusual nodes and measure the trustworthiness of nodes according to the received feedback [5]. Figure 1 presents a typical challenge-based CIDN with major components of an IDS node.

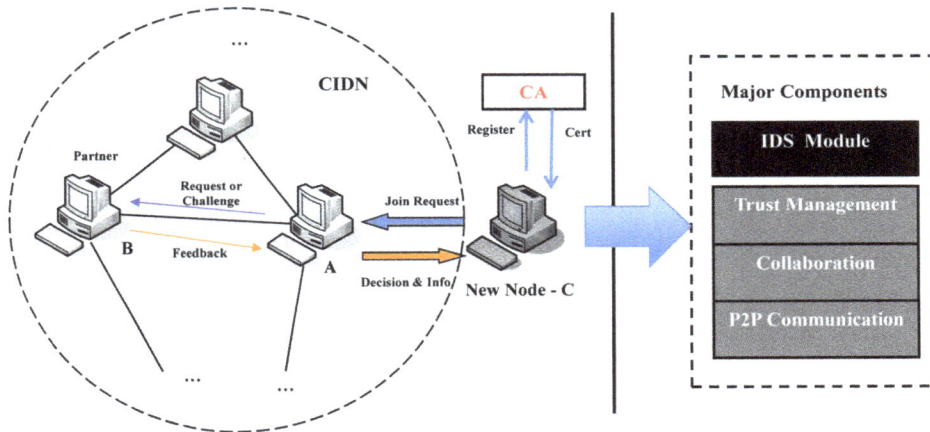

Figure 1. The high-level architecture of a typical challenge-based collaborative intrusion detection network (CIDN).

In such networks, IDS nodes can choose their own collaborators or partners in terms of their prior experience, as well as maintain a list of connected partners. This list is known as a *partner list* (or *acquaintance list*), and it can gather necessary information with other IDS nodes, for example, public keys and reputation levels. Supposing an outside node plans to join the network, it has to firstly obtain its proof of identify by registering via a trusted certificate authority (CA), including a public and private key pair. As shown in Figure 1, if node *C* plans to join the CIDN, it can apply to a node within the network, for example, node *A*. After receiving a request, node *A* can make decisions on the basis of the predefined rules and return a list of initial partners if the request is confirmed.

Interactions

To improve the detection accuracy of a separate IDS node, collaborative networks enable many IDS nodes to exchange data with other nodes; that is, several nodes can exchange alarm information to obtain a high-level view of the network status. In a challenge-based CIDN, two types of messages would be used during node interactions.

- *Challenges.* This type of message contains several IDS alarms requesting the target node to rank the severity. For instance, a testing node can send a challenge periodically to one or several tested nodes and then obtain their answers. Because the testing node extracts IDS alarms from its own database, it can know the alarm severity in advance. Accordingly, it can evaluate the tested nodes' trustworthiness by identifying the deviation between the expected and the received feedback. For the satisfaction mapping, we refer to Section 5.

- *Normal requests.* This type of message is sent by a detector to collect data for alarm aggregation. In a CIDN, if a node starts to aggregate alarms, it can send a normal request to other IDS nodes. Then, other trusted nodes can give alarm information on the basis of their own experience. Intuitively, alarm aggregation is a very important step to improve the detection accuracy of a separate intrusion detector. It is worth noting that the alarm aggregation process only considers the information from trusted nodes.

Major Components

As shown in Figure 1, an IDS node contains an *IDS module* and consists of three major components, the *trust management component*, *collaboration component* and *P2P communication*.

- *Trust management component.* To measure the reputation of IDS nodes, this component is responsible for comparing the expected answer with the received feedback. As mentioned above, each IDS node can request for the alarm severity through sending either normal requests or challenges. In order to protect challenges, Fung et al. [5] assumed that challenges should be delivered randomly, associated with timing, making them hard to be identified from a normal request.
- *Collaboration component.* The goal of this component is to handle CIDN messages, that is, to help a node measure the reputation of others by sending normal requests or challenges. For example, this component can return the answers when an IDS node receives a CIDN message. In Figure 1, node *B* would return its feedback according to its own experience, if node *A* delivers a request or a challenge.
- *P2P communication.* This component aims to help maintain connections with other nodes, that is, by configuring the network initialization, address management and node-to-node communication. The trust of the P2P communication is assumed to be trusted.

Robustness

A line of research studies (e.g., [5–7]) have shown that challenge-based trust mechanisms can protect CIDNs against common threats such as a sybil attack, a newcomer (re-entry) attack, and a betrayal attack.

- *Sybil attack.* This kind of attack describes the situation in which a node tries to create many fake identities [25]. These fake identities can be utilized to gain a larger impact on alarm aggregation in a CIDN node. The challenge mechanism mitigates this attack through requesting each IDS node to register via a trusted CA and obtain a unique proof identity.
- *Newcomer (re-entry) attack.* This type of attack indicates the situation in which a node tries to re-enter the network as a newcomer, aiming to erase its bad history. The challenge mechanism avoids this attack by allocating a low reputation level to all new joined nodes.
- *Betrayal attack.* This kind of attack indicates the situation in which a trusted node turns out to be an untruthful node unexpectedly. The challenge mechanism mitigates this attack by employing a strategy: that is, a high reputation level can only be achieved after a long time-period of interaction with consistent good behavior, whereas the reputation can be quickly degraded by detecting only a few bad actions. To realize this strategy, a forgetting factor can be used to give more weight to recent behavioral events.

Overall, challenge-based CIDNs can encourage collaborations among various IDS nodes, as well as identify common insider attacks. However, it is found that challenge-based CIDNs would suffer from advanced insider attacks, because the adopted assumptions are not realistic in real-world implementations [8,20,21].

3.2. Assumption Analysis

Challenge-based CIDNs are shown to be robust against several common insider attacks in prior studies [6,7]. However, challenge mechanisms rely on two major assumptions, causing CIDNs to be problematic in real implementations.

- *First assumption.* Challenges should be sent randomly to ensure they are hard to be identified from normal messages.
- *Second assumption.* Malicious nodes always behave untruthfully to other nodes, that is, by sending untruthful answers to the received messages.

The first assumption indicates that an IDS node cannot distinguish a challenge from normal messages, ensuring that it has a small possibility for a malicious node to give manipulated feedback to challenges. However, this assumption still leaves a chance for attackers to figure out the challenges in practice. In literature, Li et al. [8] designed an advanced attack, which can distinguish normal requests from messages with a high probability and enable nodes to send untruthful feedback only to normal requests without decreasing their trust levels.

The second assumption attempts to ensure that malicious nodes always behave abnormally, which helps to decrease their reputation in a fast manner. Fung et al. [5] summarized this assumption as a *maximal harm model*, in which a malicious node chooses to give untruthful answers with the purpose of making the most negative influence on the target nodes. However, in a practical scenario, a malicious node can choose a different harm model and send malicious answers in a dynamic way to keep the trust level.

Discussion. The above assumptions are realistic in some scenarios in which an intruder is naive and willing to use the maximal harm model. However, many intruders may choose a different and dynamic strategy to attack CIDN nodes. For instance, a malicious insider node can give untruthful feedback to some nodes while behaving normally to other nodes. On the whole, the assumptions adopted by existing challenge mechanisms are too strong for real-world implementations, leaving a chance for advanced attackers to compromise the CIDN security.

4. SOOA: Special On–Off Attack

On the basis of the above analysis, challenge-based CIDNs may suffer from advanced attacks in real scenarios. In this work, our motivation is to explore the impact of a SOOA on the robustness of challenge mechanisms. Our attack allows a node to respond truthfully to certain nodes but behave maliciously to the other nodes. It is worth emphasizing that we only accept the first assumption but improve the second assumption: that is, a malicious node can deliver untruthful answers with a strategy.

On–Off Attacks

In literature, a general on–off attack indicates the situation in which an attacker behaves well and badly alternately, aiming to compromise the network if they remain as trusted nodes [26]. The type of attack has two major states: *on-state*, when the associated action is effectively happening, and *off-state*, when the associated action is not happening. By behaving as a good node and as a bad node alternately, this type of attack pretends to be a temporary error for a security mechanism. A balance is often made: that is, a high ratio of off-state in relation to on-state is a more effective attack, while a low ratio might make it more easy for a trust management scheme to detect the malicious behavior.

Special On–Off Attack

As a study, this work considers a SOOA, for which a malicious node can keep sending truthful answers to one node but behave maliciously to another node. This particular attack has the potential to affect the effectiveness of the trust computation for a third node (target node). This attack accepts that a challenge could be delivered randomly to make it difficult to be identified from normal messages. The sending randomness can be achieved by a random number generator. Figure 2 describes an example of the SOOA: supposing node D is malicious, node A is the target, and node B and node C are partner nodes of node A. Two adversary scenarios are described below.

- **Scenario 1: node D is not a partner node of node A.** In this scenario, node D chooses to send a truthful response to node C while sending untruthful (or malicious) answers to node B. Figure 2 shows that node A can communicate and collect data with/from its partner nodes; thus, node A may receive different (or opposite) reports on node D. This scenario often occurs for a hierarchical network structure, for which a central server needs to collect information and judge the trustworthiness of each node.
- **Scenario 2: node D is a partner node of node A.** In this scenario, node D can respond truthfully to node A if they are partner nodes. In CIDNs, node A can judge the trustworthiness of node D through both its own trust computation and the judgement from other nodes, for example, nodes B and C. In this case, node D can perform the same as in Scenario 1 to affect the decision, such as by alarm aggregation of node A.

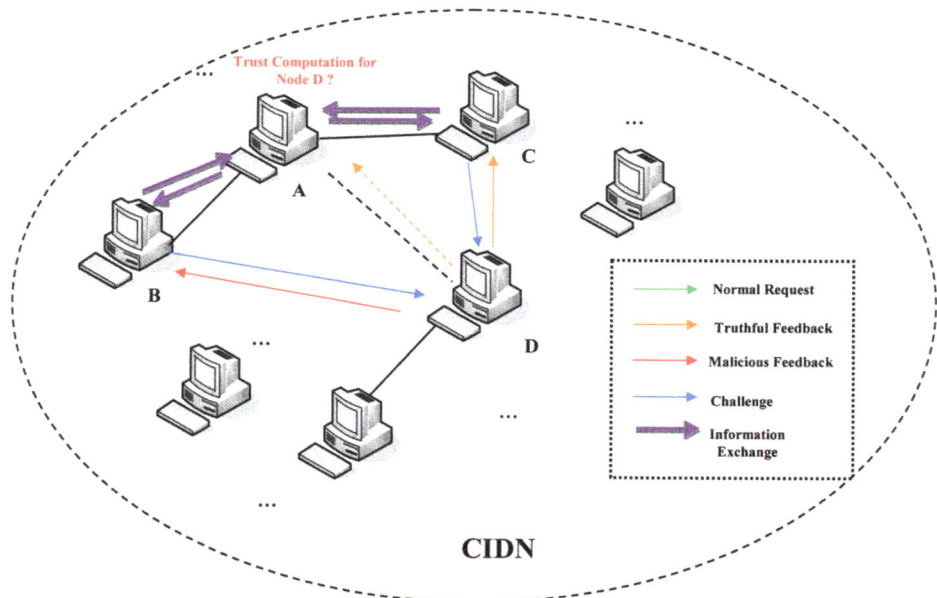

Figure 2. An example of a special on–off attack (SOOA) on challenge-based collaborative intrusion detection networks (CIDNs).

On the whole, our SOOA can choose to give truthful feedback to several nodes while responding untruthfully to others. As a result, it may affect the trust computation of certain nodes and maintain its trust values above the threshold, that is, maintaining the trust values of node D above the threshold as for node A. In practice, malicious nodes can make a negative impact on alarm aggregation in both scenarios, through maintaining their trust values.

5. Evaluation

In this section, we measure the performance of the SOOA in both a simulated and a real CIDN environment, under either Scenario 1 or Scenario 2, respectively. In the remaining parts, we describe the deployment of CIDNs and the calculation of reputation levels, as well as discuss the obtained results.

5.1. CIDN Settings

In the simulated CIDN environment, a total of 15 nodes were distributed randomly within a 5×5 grid region. Each node employs an open-source IDS plugin (Snort [27]) and builds a partner list after connecting to other IDS nodes. Similarly to a previous work [7], we chose the initial trust levels of all nodes to be $T_s = 0.5$.

To measure the reputation levels of partner nodes, each node can deliver challenges in a random manner with an average rate of ε. Two request frequencies are considered in this work: ε_l and ε_h. The request frequency would be at a low level for highly trusted or highly untrusted nodes, because their reputation levels are relatively stable. In contrast, the request frequency would be at a high level for the nodes whose trust levels are close to the detection threshold. In this work, we selected the low request frequency to be 10 per day, which was more strict than for the previous work (e.g., [7]). Table 1 shows the detailed parameters for the deployed CIDN.

Table 1. Simulation parameters in the experiment.

Parameters	Value	Description
λ	0.9	Forgetting factor
ε_l	10/day	Low request frequency
ε_h	20/day	High request frequency
r	0.8	Trust threshold
T_s	0.5	Trust value for newcomers
m	10	Lower limit of received feedback
d	0.3	Severity of punishment

Node Expertise

Similarly to former studies, this work also considered three expertise levels for a CIDN node: low (0.1), medium (0.5) and high (0.95). The IDS's expertise can be modeled by means of a beta function as shown below:

$$f(p'|\alpha, \beta) = \frac{1}{B(\alpha, \beta)} p'^{\alpha-1}(1 - p')^{\beta-1}$$
$$B(\alpha, \beta) = \int_0^1 t^{\alpha-1}(1 - t)^{\beta-1}dt \tag{1}$$

where $p'(\in [0, 1])$ describes the probability of an IDS in examining an intrusion, and $f(p'|\alpha, \beta)$ describes the probability that an IDS node at the expertise level l responds in p' to an intrusion examination of difficulty level $d(\in [0, 1])$. A higher l value means a larger probability of correctly detecting an intrusion, while a higher d value means that an intrusion is harder to identify. Further, α and β can be set as below:

$$\alpha = 1 + \frac{l(1 - d)}{d(1 - l)}r$$
$$\beta = 1 + \frac{l(1 - d)}{d(1 - l)}(1 - r) \tag{2}$$

where $r \in \{0,1\}$ is the expected result of the detection. Regarding a fixed difficulty level $d(\in [0,1])$, the node with higher expertise should have a larger probability of correctly identifying an intrusion. For instance, a node with an expertise level of 1 can accurately identify an intrusion if the difficulty level is 0.

Node Trust Evaluation

An IDS node can send a challenge periodically to measure the reputation of a tested node. A satisfaction level can be derived on the basis of the difference between the expected answers and the received feedback. As a result, the trustworthiness of a node i according to node j can be computed as below:

$$T_i^j = (w_s \frac{\sum_{k=0}^n F_k^{j,i} \lambda^{tk}}{\sum_{k=0}^n \lambda^{tk}} - T_s)(1-x)^d + T_s \tag{3}$$

where $F_k^{j,i} \in [0,1]$ represents the satisfaction level of a received feedback k, n is the total number of received feedbacks, λ is the *forgetting factor* by allocating more weight to recent answers, and w_s is the *significant weight*, which relies on the number of received feedbacks. If the number of received feedbacks is below a minimum threshold m, then $w_s = \frac{\sum_{k=0}^n \lambda^{tk}}{m}$; otherwise $w_s = 1$; x is the percentage of "don't know" replies for a period of time; d is a positive incentive parameter, which is used to control the severity of the punishment to "don't know" replies. For the detailed equation derivation, we refer to [6,7].

Satisfaction Evaluation

We let $(e \in [0,1])$ denote an expected feedback and $(r \in [0,1])$ denote a true received feedback. A function $F (\in [0,1])$ can be defined to calculate the satisfaction level by identifying the difference between the the expected feedback and the received feedback [7]:

$$F = 1 - (\frac{e-r}{max(c_1 e, 1-e)})^{c_2} \quad e > r \tag{4}$$

$$F = 1 - (\frac{c_1(r-e)}{max(c_1 e, 1-e)})^{c_2} \quad e \le r \tag{5}$$

where c_1 and c_2 control the degree of penalty for wrong estimates and the satisfaction sensitivity, respectively. Similarly to a former study [7], this work sets $c_1 = 1.5$ and $c_2 = 1$ in the evaluation.

5.2. Simulation Experiment

In this experiment, we aimed to investigate the influence of the SOOA on challenge-based CIDNs under both Scenario 1 and Scenario 2. Following the example given in Figure 2, we set up two scenarios as below:

- **Scenario 1.** In this condition, node A had six partner nodes in its list without node D. In this case, node D behaved normally to some partner nodes of node A, while it behaved untruthfully to the remaining partner nodes.
- **Scenario 2.** In this scenario, node A had seven partner nodes including node D, which could send truthful answers to several partner nodes of node A, while sending untruthful (or malicious) answers to the rest of the partner nodes.

Figure 3 depicts the convergence of trust values for different expert nodes: low ($I = 0.1$), medium ($I = 0.5$) and high ($I = 0.95$). The results validated the observations obtained in previous studies [6,7]: that is, nodes with higher expertise can achieve bigger trust values. The trust values of all nodes became stable after around 16 days in the simulated network.

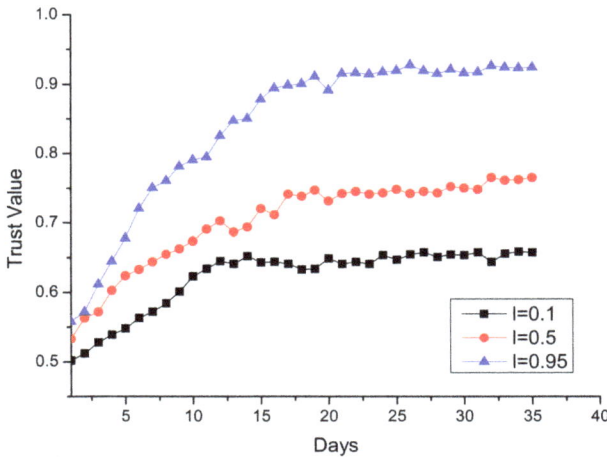

Figure 3. Convergence of trust values of intrusion detection system (IDS) nodes regarding three expertise levels.

Results under Scenario 1

In this condition, we randomly selected one expert node ($I = 0.95$) to perform our attack from day 45. As node D was not a partner of node A, node A could only evaluate the trustworthiness of node D by collecting the data from its partner list. As a study, we tested two conditions: (1) node D sent truthful answers to four partner nodes of node A, but behaved untruthfully to the other two partner nodes (T4U2); (2) node D responded truthfully to three partner nodes of node A, while sending untruthful answers to the remaining partner nodes (T3U3). Figure 4 describes the trust value of the malicious node.

- *T4U2.* Under this condition, it was found that the trust value of the malicious node (node D) could gradually decrease below the threshold after nearly 15 days, while the trust value could return above the threshold over a period of time. This was because up to four partner nodes reported a "benign" status for node D. Supposing node A was a central server in a CIDN, node D could still make an impact on its judgement as long as its trust value was higher than the threshold of 0.8.
- *T3U3.* In this condition, node D could behave untruthfully to three partner nodes of node A. It was identified that the trust value of node D could keep decreasing at most cases and fall below the threshold after 15 days without going up to the threshold again. Intuitively, the detection accuracy was better than for the first condition, as one more node would report a "malicious" status for node D.

The results demonstrate that our attack has the potential to compromise the robustness of the challenge mechanism in this scenario. If behaving normally to most partner nodes, a malicious node can maintain its trust value close to the threshold.

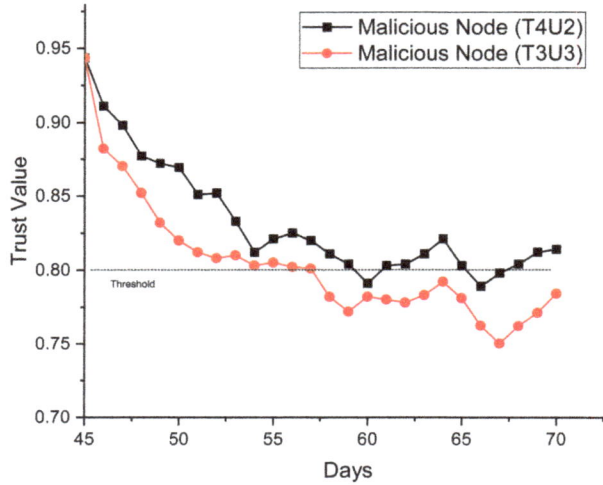

Figure 4. Trust values of malicious node (e.g., node D) calculated by target node (e.g., node A) under Scenario 1.

Results under Scenario 2

In this scenario, we randomly selected one expert node ($I = 0.95$) as a malicious node (e.g., node D), which conducted our special attack from day 45. Node D had to behave truthfully to node A, as node A could directly send challenges to node D. The trust computation of target node (node A) regarding node D is shown in Figure 5. We also tested two conditions: (1) node D behaving truthfully to four partner nodes of node A, whereas it provided untruthful feedback to the remaining two partner nodes (T4U2); (2) node D responding truthfully to three partner nodes but giving untruthful answers to the remaining three partner nodes (T3U3). It is worth noting that node D always sent truthful feedback to node A. The trust value of the malicious node is shown in Figure 5.

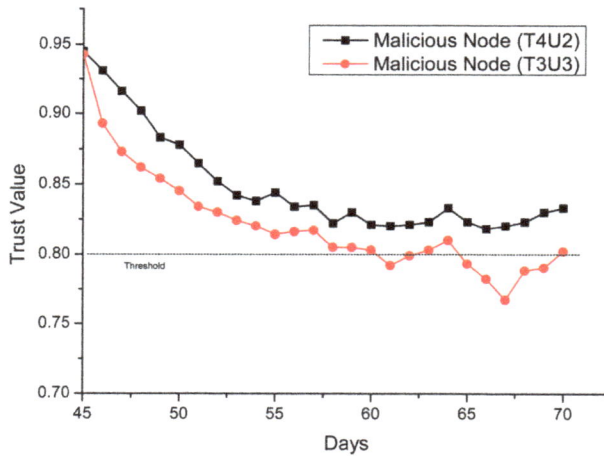

Figure 5. Trust values of malicious node (e.g., node D) calculated by target node (e.g., node A) under Scenario 2.

- *T4U2.* In this condition, the trust value of node *D* computed by node *A* could gradually decrease closer to the threshold during the first 10 days, because two partner nodes could report malicious actions of node *D* to node *A*. Afterwards, the trust value was maintained in the range from 0.81 to 0.82 in most cases, as four partner nodes reported that node *D* was normal. As the trust value was higher than the threshold of 0.8, node *D* could still make an influence on node *A* and its alarm aggregation.

- *T3U3.* In this condition, node *D* sent truthful feedback to three partner nodes of node *A* but sent malicious feedback to the other three partner nodes. It was found that the trust value of node *D* computed by node *A* could keep decreasing during the first 15 days, whereupon it was maintained around the threshold. As node *D* always sent truthful feedback to node *A*, its trust value crossed below and above the threshold.

As compared to the results reported in [6,7], the experimental results showed that our attack could greatly degrade the effectiveness and robustness of challenge-based CIDNs, for which a malicious node has a non-trivial chance to maintain its trust value above the threshold and affect the alarm aggregation for a target node. Further, we noticed that the trust value of a malicious node decreased faster in Scenario 1 than in Scenario 2. This was because node *D* could send truthful answers to node *A* in Scenario 2, as they were partner nodes, while in Scenario 1, node *A* could only evaluate the trustworthiness of node *D* on the basis of the feedback from its partner nodes.

5.3. Evaluation in a Real Environment

To explore the practical performance of the SOOA, we collaborated with an IT company and conducted a real evaluation in a wired CIDN including 26 nodes. Figure 6 shows the high-level network deployment. Each node could access the Internet by connecting to a server, which acted similarly to a firewall and provided various computing resources.

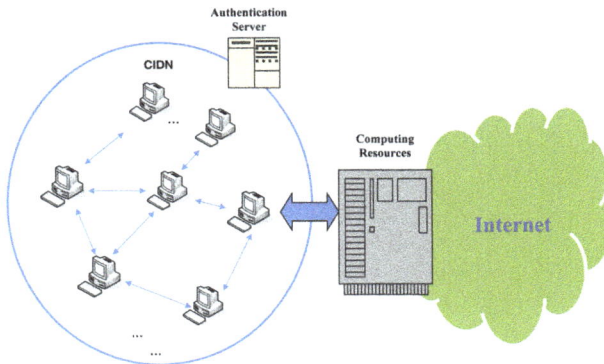

Figure 6. The high-level architecture of a real collaborative intrusion detection network (CIDN) environment.

To validate the results obtained in the simulated environment, we adopted the same environmental settings and observed the network to become stable; that is, the trust values turned out to be stable. We then set up the same scenarios and randomly selected one expert node to launch our attack. Figure 7 depicts the trust value of the malicious node under both Scenario 1 and Scenario 2.

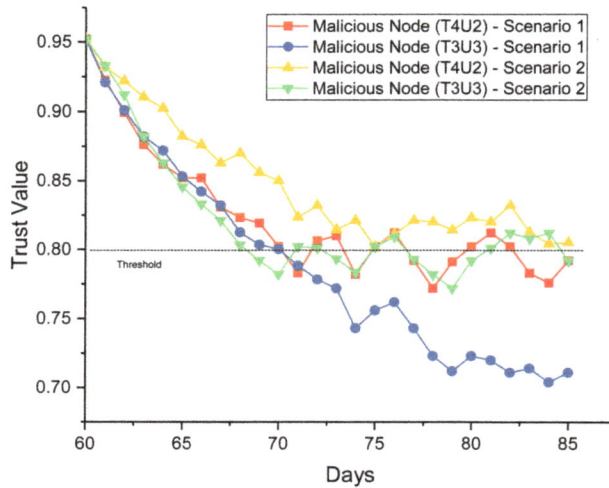

Figure 7. Trust value of malicious node under different scenarios in a real environment.

- *Scenario 1.* In this scenario, it was found that the trust value of the malicious node could decrease much faster under T3U3 than T4U2. More specifically, the trust value under T4U2 could deviate around the threshold, while decreasing nearly directly under T3U3. The observations are in-line with the results obtained in the simulated environment. While we identified that the trust value of the malicious node decreased faster in a real network environment under T3U3, an attacker should make a better strategy to launch the SOOA, that is, behave normally to more than half the nodes.

- *Scenario 2.* Figure 7 shows that the trust value of the malicious node under T3U3 could maintain a downtrend for the first 10 days but would deviate around the threshold later; that is, it was above the threshold for a total of 10 days. Under T4U2, it was found that the trust value of the malicious node would not decrease below the threshold. Similarly, the observations are in-line with the results obtained in the simulated evaluation.

On the whole, the experimental results in the real CIDN environment validated that our SOOA has the potential to compromise the robustness of the challenge mechanism by slowing the detection speed of malicious nodes. In Scenario 2, malicious nodes may have a greater chance to bypass the detection than in Scenario 1, as they can pretend to be normal to the target node directly. Furthermore, our attack could help to maintain the trust values of malicious nodes by employing a good strategy, for example, T4U2, behaving normally to most partners of the target node.

5.4. Discussion

The above experimental results have demonstrated the potential of the SOOA in compromising the robustness of challenge-based CIDNs, that is, slowing the detection speed of malicious nodes. As this is an initial study in this field, there are some improvements that can be made in our future work.

- *The impact of partner nodes' selection.* In this work, we assume that each partner has the same impact on the decision in the target node. Thus, there is no need to consider how to select a partner node for the SOOA. However, if we consider that the target node may give different weights to its partner nodes, then there is a need to identify which type of partner node can be attacked. This is an interesting topic for our future work.

- *A variety of combinations.* In this work, we only evaluated two combinations, namely, T3U3 and T4U2, under two scenarios. The obtained results have demonstrated the influence of the SOOA on the robustness of the challenge mechanisms. In future work, it will be an interesting topic to investigate the trend of trust values in other combinations, for example, T4U3 and T5U5.

To defend against this type of attack, there is a need to consider deploying additional security tools to enhance the performance of challenge mechanisms.

- *Emphasizing the impact of malicious actions.* In CIDNs, if a node wishes to evaluate a node's trustworthiness, it has to collect information from other trusted nodes. Our attack allows malicious nodes to behave maliciously, without a timely detection. One potential solution is to punish more for malicious actions, if detected by a node, for example, by building a trust computation by means of IS, which can give greater weight to expert nodes.
- *Employing additional measurements.* In challenge-based CIDNs, the trustworthiness of a node is mainly determined by challenges, but the challenges have to be sent over a period of time, rendering the network vulnerable to advanced attacks. To enhance the robustness of CIDNs, additional measures should be considered to evaluate the trustworthiness of a node, for example, packet-level trust [11].

Overall, our work validates that existing challenge mechanisms would suffer from advanced insider attacks in practice, as a result of the adopted assumptions. Additional security mechanisms can be considered to enhance the robustness of challenge-based CIDNs.

6. Conclusions

Challenge mechanisms are proved to be robust to common insider attacks, which evaluate the trustworthiness of others by sending challenges. However, we found that such kinds of mechanisms may still suffer from advanced insider attacks, as a result of the adopted assumptions. In this work, we develop a SOOA, which can behave normally to one or several nodes while sending untruthful answers to other nodes. To explore the attack performance, we performed two major experiments under both simulated and real CIDN environments. The experimental results indicate that our attack enables insider nodes to behave maliciously without being detected timely. Future work will include exploring how to enhance the existing framework to protect against advanced insider attacks, that is, applying the concept of IS.

Author Contributions: W. Li and W. Meng initialized the idea, performed the experiments and analyzed the data; L.F. Kwok designed the experiments and contributed to the paper writing.

Conflicts of Interest: The authors declare no conflict of interest.

Abbreviations

The following abbreviations are used in this manuscript:

IDS   Intrusion detection system
CIDN   Collaborative intrusion detection network
SOOA   Special on–off attack
IS   Intrusion sensitivity
PMFA   Passive message fingerprint attack

References

1. Scarfone, K.; Mell, P. *Guide to Intrusion Detection and Prevention Systems (IDPS)*; NIST Special Publication: Gaithersburg, MD, USA, 2007; pp. 800–894.
2. Gong, F. *Next Generation Intrusion Detection Systems (IDS)*; McAfee Network Security Technologies Group: Santa Clara, CA, USA, 2003.

3. Duma, C.; Karresand, M.; Shahmehri, N.; Caronni, G. A Trust-Aware, P2P-Based Overlay for Intrusion Detection. In Proceedings of the 17th International Workshop on Database and Expert Systems Applications, Krakow, Poland, 4–8 September 2006; pp. 692–697.

4. Wu, Y.-S.; Foo, B.; Mei, Y.; Bagchi, S. Collaborative Intrusion Detection System (CIDS): A Framework for Accurate and Efficient IDS. In Proceedings of the 2003 Annual Computer Security Applications Conference (ACSAC), Las Vegas, NV, USA, 8–12 December 2003; pp. 234–244.

5. Fung, C.J.; Boutaba, R. Design and management of collaborative intrusion detection networks. In Proceedings of the 2013 IFIP/IEEE International Symposium on Integrated Network Management (IM), Ghent, Belgium, 27–31 May 2013; pp. 955–961.

6. Fung, C.J.; Baysal, O.; Zhang, J.; Aib, I.; Boutaba, R. Trust Management for Host-Based Collaborative Intrusion Detection. In *DSOM 2008, Lecture Notes in Computer Science (LNCS) 5273*; De Turck, F., Kellerer, W., Kormentzas, G., Eds.; Springer: Berlin, Germany, 2008; pp. 109–122.

7. Fung, C.J.; Zhang, J.; Aib, I.; Boutaba, R. Robust and scalable trust management for collaborative intrusion detection. In Proceedings of the 11th IFIP/IEEE International Conference on Symposium on Integrated Network Management (IM), Long Island, NY, USA, 1–5 June 2009; pp. 33–40.

8. Li, W.; Meng, Y.; Kwok, L.-F.; Ip, H.H.S. PMFA: Toward Passive Message Fingerprint Attacks on Challenge-based Collaborative Intrusion Detection Networks. In Proceedings of the 10th International Conference on Network and System Security (NSS), Taipei, Taiwan, 28–30 September 2016; pp. 433–449.

9. Meng, M.; Luo, X.; Li, W.; Li, Y. Design and Evaluation of Advanced Collusion Attacks on Collaborative Intrusion Detection Networks in Practice. In Proceedings of the 15th IEEE International Conference on Trust, Security and Privacy in Computing and Communications (TrustCom), Tianjin, China, 23–26 August 2016; pp. 1061–1068.

10. Li, W.; Meng, Y.; Kwok, L.-F. SOOA: Exploring Special On-Off Attacks on Challenge-Based Collaborative Intrusion Detection Networks. In Proceedings of the 12th International Conference on Green, Pervasive, and Cloud Computing (GPC), Cetara, Italy, 11–14 May 2017; pp. 402–415.

11. Meng, Y.; Li, W.; Xiang, Y.; Choo, K.-K.R. A Bayesian Inference-based Detection Mechanism to Defend Medical Smartphone Networks against Insider Attacks. *J. Netw. Comput. Appl.* **2017**, *78*, 162–169.

12. Donovan, S.; Feamster, N. Alternative trust sources: Reducing DNSSEC signature verification operations with TLS. In Proceedings of the 2015 ACM Conference on Special Interest Group on Data Communication (SIGCOMM), London, UK, 17–21 August 2015; pp. 353–354.

13. Kolias, C.; Kolias, V.; Kambourakis, G. TermID: A distributed swarm intelligence-based approach for wireless intrusion detection. *Int. J. Inf. Secur.* **2017**, *16*, 401–416.

14. Meng, Y.; Li, W.; Kwok, L.-F. Evaluation of detecting malicious nodes using Bayesian model in wireless intrusion detection. In Proceedings of the 7th International Conference on Network and System Security (NSS), Helsinki, Finland, 21–23 August 2013; Lecture Notes in Computer Science 7873; Springer: Berlin, Germany, 2013; pp. 40–53.

15. Meng, W.; Au, M.H. Towards statistical trust computation for medical smartphone networks based on behavioral profiling. In Proceedings of the 11th IFIP WG 11.11 International Conference on Trust Management (IFIPTM), Gothenburg, Sweden, 12–16 June 2017; pp. 152–159.

16. Meng, W.; Li, W.; Wang, Y.; Au, M.H. Detecting malicious nodes in medical smartphone networks through euclidean distance-based behavioral profiling. In Proceedings of the 9th International Symposium on Cyberspace Safety and Security (CSS), Xi'an, China, 23–25 October 2017; pp. 163–175.

17. Meng, W.; Li, W.; Su, C.; Zhou, J.; Lu, R. Enhancing Trust Management for Wireless Intrusion Detection via Traffic Sampling in the Era of Big Data. *IEEE Access* **2017**, doi:10.1109/ACCESS.2017.2772294.

18. Ramos, A.; Lazar, M.; Filho, R.H.; Rodrigues, J.J.P.C. A security metric for the evaluation of collaborative intrusion detection systems in wireless sensor networks. In Proceedings of the IEEE International Conference on Communications (ICC), Paris, France, 21–25 May 2017.

19. Li, W.; Meng, Y.; Kwok, L.-F.; Ip, H.H.S. Developing Advanced Fingerprint Attacks on Challenge-based Collaborative Intrusion Detection Networks. *Clust. Comput.* **2017**, 1–12, doi:10.1007/s10586-017-0955-8.

20. Li, W.; Meng, Y.; Kwok, L.-F. Enhancing Trust Evaluation Using Intrusion Sensitivity in Collaborative Intrusion Detection Networks: Feasibility and Challenges. In Proceedings of the 9th International Conference on Computational Intelligence and Security (CIS), Leshan, China, 14–15 December 2013; pp. 518–522.

21. Li, W.; Meng, W.; Kwok, L.-F. Design of intrusion sensitivity-based trust management model for collaborative intrusion detection networks. In *Trust Management VIII, IFIP AICT*; Zhou, J., Gal-Oz, N., Zhang, J., Gudes, E., Eds.; Springer: Heidelberg, Germany, 2014; Volume 430, pp. 61–76.

22. Li, W.; Meng, Y.; Kwok, L.-F.; Ip, H.H.S. Enhancing Collaborative Intrusion Detection Networks Against Insider Attacks Using Supervised Intrusion Sensitivity-Based Trust Management Model. *J. Netw. Comput. Appl.* **2017**, *77*, 135–145.

23. Li, W.; Meng, W. Enhancing collaborative intrusion detection networks using intrusion sensitivity in detecting pollution attacks. *Inf. Comput. Secur.* **2016**, *24*, 265–276.

24. Cho, J.-H.; Chan, K.; Adali, S. A Survey on Trust Modeling. *ACM Comput. Surv.* **2015**, *48*, doi:10.1145/2815595.

25. Douceur, J. The sybil attack. In *IPTPS 2002. LNCS*; Druschel, P., Kaashoek, M.F., Rowstron, A., Eds.; Springer: Heidelberg, Germany, 2002; Volume 2429.

26. Perrone, L.P.; Nelson, S.C. A Study of On-Off Attack Models for Wireless Ad Hoc Networks. In Proceedings of the 2006 Workshop on Operator-Assisted Community Networks, Berlin, Germany, 18–19 September 2006; pp. 1–10.

27. Snort: An Open Source Network Intrusion Prevention and Detection System (IDS/IPS). Homepage. Available online: http://www.snort.org/ (accessed on 8 January 2018).

future internet

MDPI

Article

A New Lightweight Watchdog-Based Algorithm for Detecting Sybil Nodes in Mobile WSNs

Rezvan Almas Shehni [1,‡], Karim Faez [2,‡], Farshad Eshghi [3,*,†,‡] and Manoochehr Kelarestaghi [3,‡]

1   Department of Computer Engineering and Information Technology, Qazvin Azad University, Qazvin 15195-34199, Iran; r.shehni@qiau.ac.ir

2   Department of Electrical Engineering, Amirkabir University of Technology, Tehran 15916-34311, Iran; kfaez@aut.ac.ir

3   Department of Electrical & Computer Engineering , Faculty of Engineering, Kharazmi University, Tehran 15719-14911, Iran; kelarestaghi@khu.ac.ir

*   Correspondence: farshade@khu.ac.ir; Tel.: +98-912-497-0182

†   Current address: Department of Electrical & Computer Engineering , Faculty of Engineering, Kharazmi University, Tehran, Iran.

‡   These authors contributed equally to this work.

Received: 15 November 2017; Accepted: 8 Decmber 2017; Published: 21 Decmber 2017

Abstract: Wide-spread deployment of Wireless Sensor Networks (WSN) necessitates special attention to security issues, amongst which Sybil attacks are the most important ones. As a core to Sybil attacks, malicious nodes try to disrupt network operations by creating several fabricated IDs. Due to energy consumption concerns in WSNs, devising detection algorithms which release the sensor nodes from high computational and communicational loads are of great importance. In this paper, a new computationally lightweight watchdog-based algorithm is proposed for detecting Sybil IDs in mobile WSNs. The proposed algorithm employs watchdog nodes for collecting detection information and a designated watchdog node for detection information processing and the final Sybil list generation. Benefiting from a newly devised co-presence state diagram and adequate detection rules, the new algorithm features low extra communication overhead, as well as a satisfactory compromise between two otherwise contradictory detection measures of performance, True Detection Rate (TDR) and False Detection Rate (FDR). Extensive simulation results illustrate the merits of the new algorithm compared to a couple of recent watchdog-based Sybil detection algorithms.

Keywords: mobile WSN; security; watchdog node; Sybil attack

1. Introduction

Nowadays we are witnessing the emergence of new WSN applications in different fields such as military, urban services, the environment, medicine, explorations and Intrusion Detection Systems (IDS). WSNs comprise a large number of small sensor nodes featuring small memory and low power. The broadcast nature of wireless and unattended operation of WSNs necessitate the implementation and improvement of security schemes [1,2].

Operation disruption in hostile wireless networks can be realized in Physical (PHY) or higher layers. In the former case, a malicious node tries to harm communication between wireless nodes by broadcasting jamming signals [3] or imposing any other kind of interference [4] whereby normal wireless nodes become unable to interpret receiving signals. in the latter case, malicious nodes try to either deceive normal wireless nodes through disseminating fake information (e.g., Sybil, wormhole, impersonation, and etc. attacks) or overwhelm/disable normal wireless nodes using, for instance, Hello flood attack [2].

In this paper, we focus on the Sybil attack which is one of the most important attacks in WSNs. In this attack, an illegal node or a legal node captured by enemy, called a "malicious node", identifies itself by releasing several fake IDs or IDs fabricated from other legal nodes. The fake IDs represent some non-existing nodes known as Sybil nodes. As a result, legal nodes think they have many legitimate neighbors. Malicious nodes can affect routing and operational protocols such as data aggregation, voting, resource allocation, misbehavior [5,6] .

In general, Sybil attack detection techniques can be categorized into centralized and decentralized methods. Centralized methods feature a central node which is responsible for node identity management. In relevant detection methods, the information in central nodes are used for Sybil attack detection. Decentralized methods make use of some pre-authenticated nodes, sometimes called watchdogs, trusted, etc., which administrate the attack detection operation. These watchdogs can be fixed or mobile.

From a networking layer stack point of view, Sybil attack detection techniques can be grouped into PHY-layer-based and upper-layers-based techniques. The former make use of the parameters of the radio signal and the second layer's node identity information. On the other hand, the latter techniques are based on the communication of the data which is formed in the upper layers. Nevertheless, the identity information is still required in upper-layer based techniques. PHY-layer-based category is further divided into location and non-location-based techniques. Location-based techniques mostly involve analysis of Received Signal Strength Indicator (RSSI), Time Difference of Arrival (TDoA), and Angle of Arrival (AoA). One of the techniques under non-location-based category, as far as we could identify, is the Radio resource testing [6]. The techniques under the upper-layers-based category are further divided to neighborhood-based [7,8], code attestation-based [9–11], authentication-based (puzzle solving technique [12] for peer-to-peer networks, Identity certificate technique [13]), and identity registration-based [14]. Figure 1 illustrates the Sybil attack detection techniques categorization.

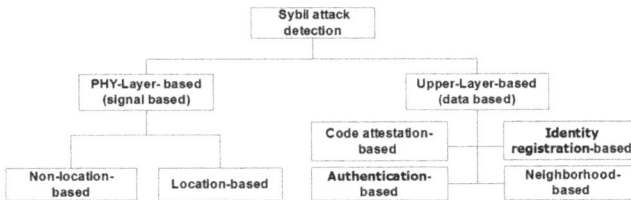

Figure 1. Categorization of Sybil attack detection algorithms in WSNs.

In mobile WSNs, Sybil nodes corresponding to a particular malicious node appear/disappear simultaneously in/from some neighborhoods. This behavior (misbehavior) can be used for Sybil nodes detection. At the same token, Watchdog-based techniques have been extensively employed to detect misbehaviors in WSNs [15,16]; thus, it is plausible to think of it as a tool for Sybil nodes detection as well. In this paper, we propose a lightweight, sufficiently accurate, and practical watchdog-based algorithm for detecting Sybil nodes in mobile WSNs. Our algorithm, considered as a neighborhood -based technique, does not need any centralized base station and does not require transmission of neighborhood data (neighborhood table) from normal nodes to watchdog nodes.

Due to being computationally lightweight, in addition to WSNs, the proposed algorithm can be well adapted to new emerging applications which involve less complex nodes such as Internet of Things (IoT), smart home/car, and health-care.

This paper is organized as follows. Section 2 discusses the related works. Problem statement is presented in Section 3. Section 4 is dedicated to the description of the proposed algorithm, followed by simulation results and performance evaluation in Section 5. Concluding remarks are drawn at the end.

2. Related Works

Sybil attack was introduced for the first time in [17] for peer- to- peer networks. In [2], it was noted that this attack can also be a dangerous threat to routing algorithms in WSNs and can be.

Newsome et al. [6] present a detailed analysis of the Sybil attack in WSNs alongside some attack detection mechanisms. Also, the taxonomy of Sybil attack are introduced in the same work which is referenced to by most researchers in the field. In what follows we review some important related works within the categorization framework introduced in Figure 1.

Focusing on PHY-layer-based approaches, in [6], several approaches in Sybil attack detection have been discussed. The first one is radio resource testing, a non-location-based technique, which relies on the assumption that a node can communicate with each of its neighbors through pre-assigned channels. When a node wants to verify one of its neighbors, it could choose a channel randomly to listen. If the neighbor that was assigned that channel is legitimate, it should hear the message. This detection technique might not work in the case of an attacker equipped with a multi-radio transceiver capable of concurrently communication through different channels.

Abbas et al. [18] propose a Sybil attack detection scheme in Mobile Ad hoc Networks (MANETs) based on monitoring and differentiating between the entry and exit RSS behavior of legitimate nodes and the Sybil attackers.

There are other techniques introduced in [6] which belong to the upper-layer-based category and will be discussed in due place. Most of Sybil attack detection approaches are based on location-verification based which use some related techniques, such as RSSI or TDOA, to distinguish between Sybil and normal nodes. Since these methods depend on some parameters of receiving signals, most probably tainted with noise and multi-path phenomena, they might end up being less reliable. The RSSI-based location determination has been used In [19] to detect Sybil nodes. Four detector nodes which are able to hear the packets from all areas of the network, cooperatively evaluate the location of the node sending the packet. Node IDs found to be sending packets from the very same location are assumed to be Sybil IDs. In [20] a solution for Sybil attack detection is proposed based on TDOA between a source node and three beacon nodes which detect the location of Sybil nodes. Since three Beacon nodes are used to calibrate the time measurements, there is some communication overhead present. Another Sybil detection system has been proposed in [21] which relies on the raging capabilities of Ultra-Wide Band (UWB) in the PHY layer. Each node periodically monitors its distance from each possible pair of its neighbors. An alarm is triggered when two or more nodes are being located in the same area. The locally computed ranging estimation is used to measure the distance between neighbors. As the authors mention, the proposed technique induces lack of compliance with old-fashioned WSNs. In this work node mobility has not been considered.

Turning our focus on upper-layer-based approaches, code attestation-based techniques are another approach mentioned in [6]. these differentiate the code running on a malicious node with that of a legitimate node [9,10]. A compromised node detection algorithm, based on code attestation, is proposed in [11] and called Unpredictable Software-based Solution (USAS). USAS administers attestation on a randomly selected nodes rather than all, in order to decrease checksum computation time. Since it only attests nodes one hop from the base station, if a compromised node is far from the base station (more than one-hop), it might not be detected.

An authentication-based approach, also introduced in [6], is a version of random key per-distribution.

In random key pre-distribution "a random set of keys or key-related information is assign to each sensor node, so that in the key set-up phase, each node can discover or compute the common keys it shares with its neighbors; the common keys will be used as a shared secret session key to ensure node-to-node secrecy". So the technique involves associating the node identity with the keys assigned to the node, and thereafter Key validation of an claimed identity. As [6] claims himself, "the problem with this approach is that if an attacker compromises multiple nodes, he can use every combination of the compromised keys to generate new identities". He presents the solution to be indirect node validation. however, it should be mentioned that indirect node validation imposes a large operational overhead on the network. Another problem with all authentication-based techniques, as is the case with this one, is the requirement of secure code pre-distribution channels which are not addressed clearly.

Another authentication-based approach is a light-weight identity certificate method which uses one-way key chains and Merkle hash trees to defeat Sybil attacks [13]. This method requires a significant amount of memory for storing information. The authors claim to have overcome this issue by means of the low level Merkle hash tree cryptographic.

The last authentication-based approaches to be reviewed, use puzzle solving techniques to detect Sybil attackers [12,22]. Another way of authentication is using puzzle-based computational mechanisms [12,22]. The main idea, herein, is that the attacker should not be able to solve a subject puzzle. There is an inherent communication overhead due to puzzle dissemination and receiving puzzle solutions back. As to the best of our knowledge, the puzzle solving techniques have not been reported in WSNs so far.

Identity registration is an centralized approach which is introduced in [6]. All nodes are registered in a trusted central authority (such as a base station) and the list of legitimate IDs are distributed amongst all nodes. To prevent the Sybil attack, each ID should be checked against the list of legitimate IDs. An important concern is that list of legitimate nodes must be protected from being maliciously modified. If the attacker is able to add IDs to this list, he will be able to add Sybil nodes to the network.

The last category of Sybil attack detection algorithms to be reviewed is the neighborhood-based methods. This category is of special interest to us since our proposed algorithm belongs to it. Ssu et al. [8] propose a neighborhood–based method for detecting Sybil nodes which uses the fact that a malicious node produces similar neighbors lists corresponding to its different Sybil IDs. Each node constructs a critical set of neighboring IDs using similarities of different neighbor lists that it receives. IDs which transmit neighbors lists containing the critical set are labeled as Sybil. In the case of mobile nodes, the neighbors lists will change so often that they result in high communicational overhead.

In [23] a Sybil attack detection method is proposed for MANETs based on cooperative monitoring. The packet receiver or forwarder can provide a proof that the sender transmitted the packet at the claimed location and time using a signature field that is unique for a non-malicious node. The results of these observations are periodically communicated between nodes. From these observations path similarities for packets originating from Sybil IDs are extracted and acted upon. The proposed algorithm involves computationally intensive procedures for signature operations, communication overhead for communicating observations, and hardware cost associated with employing directional antennas. Therefore, it seems not to be suitable for WSNs.

Another neighborhood-based method which can serve as a comparison basis for our proposed algorithm is stated in [24]. In [24] a Sybil attack detection method is proposed based on observed transmissions. This method, called Passive Ad hoc Sybil Identity Detection (PASID), uses the fact that all Sybil IDs of a single malicious node must move together because they are bound to a single physical node. Therefore, the Sybil nodes could be detected by periodically observing the network. It assumes that there is a single malicious node in the network that fabricates IDs. A subset of the legitimate nodes observe all received transmissions over time intervals. Then, the observing nodes exchange their information to identity the nodes which were heard simultaneously in the same duration. Finally, a graph-based profile of co-heard nodes is constructed where the weighted edge between two vertexes denotes their affinity. Piero et al. [24] do not study scenarios with multiple malicious nodes.

Finally, in [25], a watchdog node labels other nodes it sees as they move around. For instance, in a 4-watchdog scenario, watchdog nodes have certain labels like 00, 01, 10, and 11. Each watchdog assigns and stores its corresponding label to nodes which appear in its neighborhood. Periodically, watchdog nodes exchange their assignment information (moving_history) with each other to update the so called bit_label of their neighbor nodes. At the end, each watchdog node detects Sybil IDs by investigating bit patterns in its bit_label. Because of the periodic information exchange between watchdog nodes, and since each watchdog node has to store the whole bit_label of its neighboring nodes, the protocol imposes a lot of communication and memory overheads. Also, an error in bit_label in one of the watchdog nodes is propagated to other watchdog nodes. At the end, it seems that labeling may non-linearly increase computational, communication, and memory overheads for an increased number of watchdog nodes.

The latter two protocols [24,25] are best fit to serve as bases for comparison against our proposed protocol due to similarity in assumptions, specifically the attack model and the mobility-based-computation considerations.

3. Problem Statement and Attack Model Assumptions

In what follows we describe the network assumptions and attack models employed. The subject network consists of two sets of nodes, normal sensor nodes (SN) which perform sensing, routing, and data aggregation, and watchdog nodes (WD) which are responsible for network monitoring and detecting Sybil IDs. It is further assumed that, in order to conceal their presence, the watchdog nodes do not send any messages while overhearing the transmissions of their neighbors.

Each node has a unique ID and is not aware of its geographical location. All nodes (normal and watchdog) have the same wireless range and move according to Random-Way-Point mobility model [7] during the network lifetime. Regarding the attack models, we have considered the "Direct, Simultaneous and Fabricated IDs" Sybil attack models as described in [6]. The subject network is insecure due to the presence of some malicious nodes (MN) that fabricate some IDs, representing Sybil IDs. the malicious nodes broadcast Hello Packets using these fake IDs. This is intended to disrupt routing operation in the network.

4. Proposed Algorithm

The following entities are the core to the proposed algorithm.

- $\mathbf{A^k_{co-prs}}$: This is an upper-triangle $H \times H$ matrix which contains the co-presence status of all node pairs at time index k. The elements of A^k_{co-prs} are in the form of $xy : x, y = 0/1$ where 0 and 1 represent absence and presence respectively (Figure 2a). In Figure 2, H refers to the total number of Sybil and normal-node IDs which is equal to:

$$H = N + M * S \tag{1}$$

where $N = |SN|$, $M = |MN|$, and S equals the number of Sybil IDs per malicious node.

- $\mathbf{C^k_{co-prs}}$: This matrix which is structurally similar to A^k_{co-prs} (upper-triangle $H \times H$) scores the co-presence of each node pair at time index k (Figure 2b) and is updated according to A^{k-1}_{co-prs}, A^k_{co-prs} and the co-presence state diagram model.

- **Co-presence state diagram model:** This diagram shows how a transition between co-presence states of ID pairs updates the elements of the C^k_{co-prs} matrix (Figure 3).

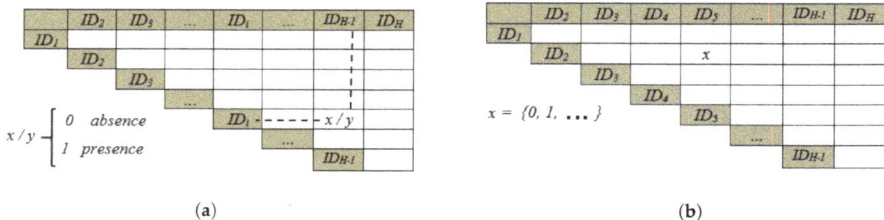

(a)   (b)

Figure 2. Structure of the WD matrices: a) A^k_{co-prs}, b) C^k_{co-prs}.

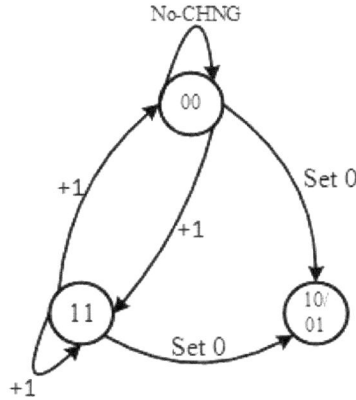

Figure 3. Co-presence state diagram model of ID pairs.

The algorithm is described in steps, with numbered references to the flowchart of Figure 4, as follows.

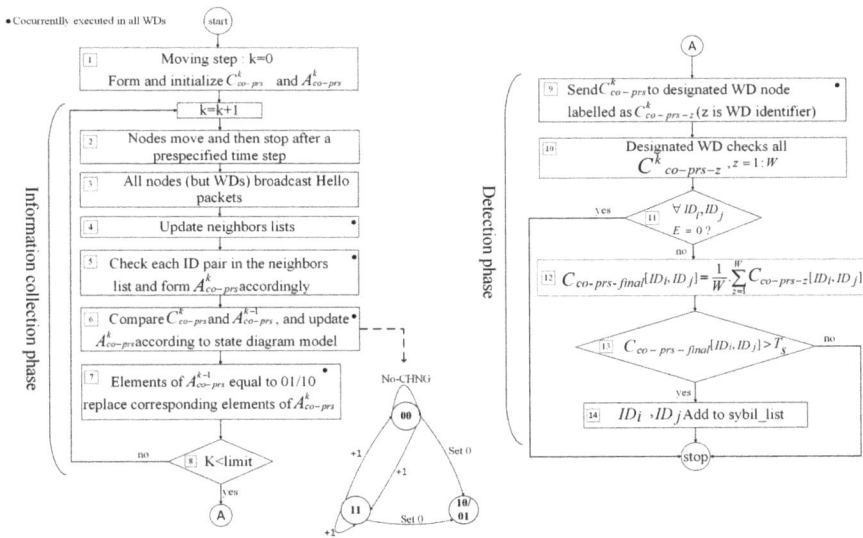

Figure 4. Flowchart of the proposed algorithm.

- **Initialization phase**

  Step I (1): Each WD constitutes of matrices A_{co-prs}^k and C_{co-prs}^k and fills the former with 00s corresponding to the general co-absence status.

- **Information collection phase**

  Step II (2 , 3 , 4): Hello Packets are broadcasted by normal and malicious nodes at fixed time intervals (movement steps). Following each movement, all WD nodes update their neighbors lists by overhearing these Hello Packets.

  Step III (5): Each WD forms the new A_{co-prs}^k based on its current neighbors list as follows.

Two-digit binary numbers 00, 11, and 10/01 correspond to co-absence, co-presence, and alternate-presence statuses respectively. Specifically, if ID_i and $ID_j \notin neighbors - list$, the content of element (ID_i, ID_j) of matrix A_{co-prs}^k is set to 00. On the other hand, if ID_i and $ID_j \in neighbors - list$, this content is set to 11. Finally, if $ID_i \in neighbors - list$ and $ID_j \notin neighbors - list$ (or vice versa), the content of the corresponding element of A_{co-prs}^k is set to 10. These values also represent the states in the co-presence state diagram model.

Step IV (6): By comparing the contents of the corresponding elements of A_{co-prs}^{k-1} and A_{co-prs}^k, the content of the corresponding element of C_{co-prs}^k is updated according to the transitions in the state diagram. For instance, when $A_{co-prs}^{k-1}[ID_i, ID_j] = 00$ and $A_{co-prs}^k[ID_i, ID_j] = 11$, according to the state diagram, $C_{co-prs}^k[ID_i, ID_j]$ must be increased.

Step V (7 , 8): In this step, A_{co-prs}^k is updated so that the elements of A_{co-prs}^{k-1} which are equal to 01/10 replace the corresponding elements of A_{co-prs}^k (trap states are preserved).

Steps II to V are repeated for a predetermined number of times representing the network's lifetime.

- **Detection phase**

Step VI (9 , 10 , 11 , 12): All WDs send their co-presence information C_{co-prs}^k (labeled as $C_{co-prs-z}^k$) to a designated WD node . The designated WD then checks the elements of $C_{co-prs-z}^k$ for each WD and creates a $C_{co-prs-final}$ matrix according to:

$$C_{co-prs-final}[ID_i, ID_j] = \frac{E.Func}{W} \tag{2}$$

wherein

$$Func = \sum_{z=1}^{W} C_{co-prs-z}^k[ID_i, ID_j] \tag{3}$$

and

$$E = \begin{cases} 0 & \text{if } \exists z \neq y \mid C_{co-prs-y}^k[ID_i, ID_j], \\ & C_{co-prs-z}^k[ID_i, ID_j] = 0, \\ & y, z = 0, 1, ..., n, \\ 1 & \text{otherwise} \end{cases} \tag{4}$$

Step VII (13 , 14): The designated WD examines the elements of $C_{co-prs-final}$ against a predetermined Sybil threshold, T_s. If $C_{co-prs-final}[ID_i, ID_j]$ is greater than T_s, ID_i and ID_j are added to its internally maintained Sybil list. T_s is a representation of how often on average two non-sybil IDs are expected to co-appear in one WD's neighborhood. T_s can be specified by trial and error to generate satisfactory true and false detection results. The designated WD finally broadcasts the Sybil list to other WDs to act upon.

Notes :

- There are rare circumstances where the proposed algorithm falsely detects normal nodes as Sybil IDs (false negative). In particular, this happens when a malicious node and a normal node simultaneously move in and out of a WD's neighborhood and because the proposed algorithm operates based on co-appearance detection .

- While the detection phase is implemented in the designated WD node, it is no different from other WDs. If the designated WD fails, provisions could be put in place (as a future work) to replace it with another WD. Thus, the algorithm can be thought of as being somewhat protected from the single-point-of-failure problem.

A Typical Example: As a typical example, assume a scenario in which there is a sensor network consisting of four normal nodes ($ID_1, ID_2, ID_6,$ and ID_7), four WDs ($W_1, W_2, W_3,$ and W_4 one of which plays the role of the designated WD), and one malicious node (labeled M, generating three Sybil IDs (namely $ID_3, ID_4,$ and ID_5) all moving around according to the Random-Way-Point movement model.

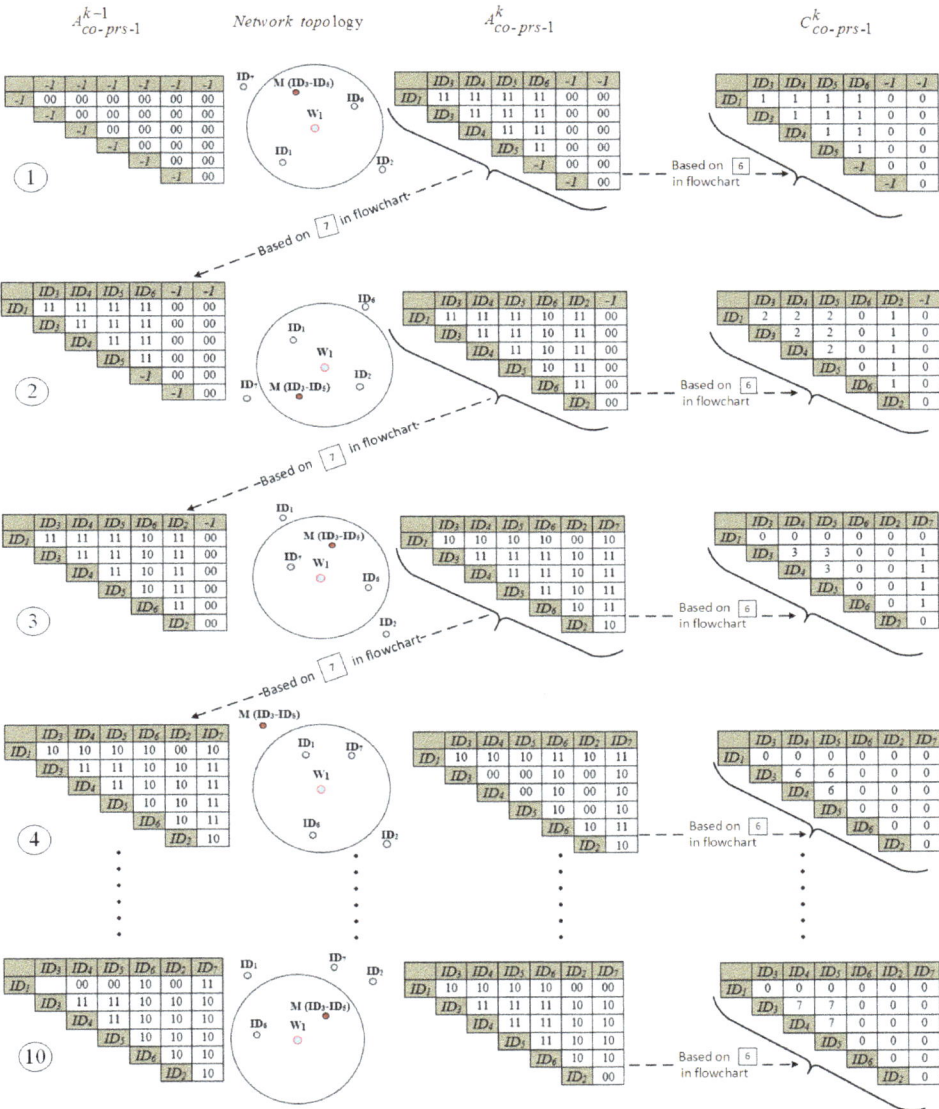

Figure 5. A pictorial representation of the advancement of the information collection phase in a typical watchdog, herein, W_1.

- Information collection phase: Figure 5 shows the information collection phase of the algorithm illustrating how it proceeds step-by-step as is implemented in W_1. The same procedure is repeated in $W_2 - W_4$ not shown here for the sake of brevity.

In each row (corresponding to a specific time step), the left most column illustrates the former IDs-co-presence status of the network shown by matrix $A_{co-prs-1}^{k-1}$. The second column from the left illustrates the network topology after applying one-step movement.

Similarly, the third column from the left shows the current (post-movement) IDs-co-presence status of the network presented by matrix $A_{co-prs-1}^{k}$. By comparing $A_{co-prs-1}^{k-1}$ and $A_{co-prs-1}^{k}$, the scoring matrix $C_{co-prs-1}^{k}$ is updated according to the state diagram of Figure 3, as shown in the forth column. This algorithm proceeds in the following row (corresponding to the next time step) starting with a $A_{co-prs-1}^{k-1}$ generated from $A_{co-prs-1}^{k}$ in the preceding row and manipulated by $\boxed{7}$ in the flowchart of Figure 4. This phase is terminated by reaching the simulation step limit which is equal to 10 in this example.

- Detection phase: The designated WD receives the $C_{co-prs-z}^{k}$ matrices from the other three WDs. Then the designated WD uses the received information, and its own $C_{co-prs-z}^{k}$ matrix to form the $C_{co-prs-final}$ matrix according to Equation (2). Finally, each element of the $C_{co-prs-final}$ matrix is compared against the Sybil threshold, $T_s = 1$, to detect and announce the Sybil IDs as illustrated in Figure 6.

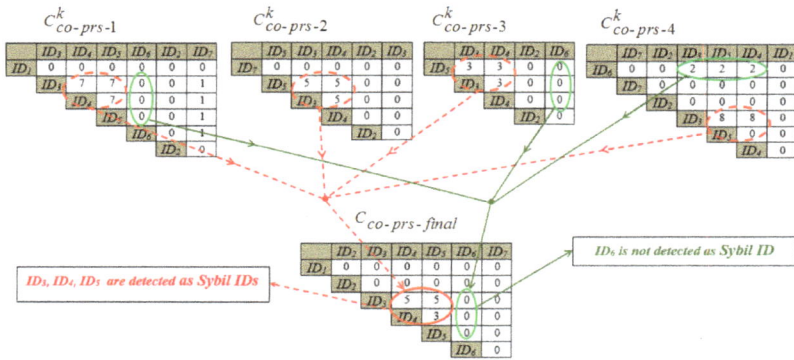

Figure 6. A pictorial representation of the advancement of the detection phase (Sybil IDs announcement) in the designated WD which is selected from $W_1 - W_4$ beforehand.

5. Simulation Results and Performance Evaluation

In this section, first we start with introducing the proper measures of performance needed for evaluating our algorithm. This is followed by describing the simulation setup. Finally, the merits of our algorithm is studied in the simulation results section.

5.1. Measures of performance

Measures of performance commonly used for evaluating the efficiency of detection algorithms are as follows.

- True Detection Rate (*TDR*): the percentage of Sybil nodes detected by a detection algorithm.
- False Detection Rate (*FDR*): the percentage of normal nodes detected as Sybil nodes erroneously.
- Memory overhead: the amount of memory consumption for algorithm implementation.
- Communicational overhead: the amount of extra algorithm-specific control-information required for algorithm implementation.
- Computational load: the number of computational operations needed for implementing an algorithm.

5.2. Simulation Setup

The proposed algorithm is simulated using J-SIM ([26]) where its physical layer employs free-space and Two-ray ground models. Regarding the topology of the network, there are a total number of N normal nodes, W watchdog nodes, and M malicious nodes all of which move according to the 2D Random-Way-Point process and with a random speed confined to a maximum speed limit. Initially, all nodes are randomly located over the network area. Each malicious node fabricates S number of Sybil IDs. The value of the above parameters and other simulation parameters appear in Table 1. Regarding the Sybil threshold, T_s, the adoption of value 1 represents an astringent choice.

Table 1. Simulation Parameters.

Parameter	Value/Fixed	Value/Variable
No. of normal nodes, N	300	$100 \rightarrow 400$
No. of watchdog nodes, W	4	$4 \rightarrow 10$
No. of Sybil IDs/malicious node, S	14	$10 \rightarrow 22$
No. of malicious nodes, M	5	$1 \rightarrow 10$
Max speed of nodes	5 m/s	-
Topology size	100 m × 100 m	-
Wireless radio range	10 m	-
Simulation time step	50 ms	-
Sybil threshold, T_s	1	-

5.3. Simulation Results

In this section, we evaluate our algorithm by setting up four tests, and thereafter we compare our algorithm with two more relevant recent works.

5.3.1. Evaluation

The TDR and FDR of the new algorithm are evaluated by setting up five tests. These tests differ by the choice of the independent and fixed parameters. To ensure the validity of the results, each point is the average of 30 repetitions to achieve a 95% confidence interval.

At the end, the three remaining measures of performance, memory/communicational overhead, and computational load are discussed qualitatively.

- **Test 1:** This test is designed to evaluate the new algorithm's TDR and FDR against the number of movement steps for different number of Sybil IDs as a parameter varying from 8 to 20. The value of the remaining parameters are fixed as they appear in Table 1.
  Figure 7a shows that TDR increases as time goes by. This is completely expected since more information is collected in longer periods of time. After 120 movement steps, for any number of Sybil IDs, good detection rates of at least 95% are achieved.
  Similarly, we expect FDR to improve by collecting more information over time towards higher movement steps. As Figure 7b illustrates, regardless of the number of Sybil IDs, FDR sufficiently nears zero at movement steps close to 160.
- **Test 2:** In this test, TDR and FDR have been evaluated against the number of movement steps for different number of normal nodes varying from 100 to 400 (The values of the remaining parameters are fixed as they appear in Table 1).

  As shown in Figure 8a and very consistent with the results of Figure 7a, good TDR results are obtained after 120 number of movement steps, no matter what the number of normal nodes is Similar to what we observed in Figure 7b of Test 1, Figure 8b shows that FDR decreases with increasing movement steps until reaching almost zero at movement steps close to 200.

- **Test 3:** Through this test it is illustrated how TDR and FDR are affected by varying the number of malicious nodes from 1 to 10. The fixed parameters are $S = 10, W = 4$, and $N = 200$ (for the values of other parameters refer to Table 1).

  In Figure 9a,b, almost perfect TDRs and FDRs are achieved for all number of malicious nodes from movement steps greater than or equal to 120 on.

- **Test 4:** Values of TDR and FDR against the number of Sybil IDs for different number of watchdog nodes is the subject of this test. The results correspond to movement step equal to 160 where the system has already reached its steady state. As before, the other fixed parameters are as they appear in Table 1.

  Figure 10a suggests that while increasing the number of WDs results in better TDRs, the improvement is negligible after some case-dependent WD population ($W = 6$ and greater). Figure 10b illustrates perfect FDR at the steady state for all choices of WD population.

- **Test 5:** In this test, the effect of scalability on the TDR/FDR performance of the proposed algorithm is verified (Figure 11). To make it a fair comparison, all node populations (normal, malicious, and watchdog) grow proportionally with the network area. The value of population parameters and the snapshot instance (Movement step) are mentioned in the figure. The maximum network size adopted is constrained by the limitations of J-SIM. The other fixed parameters are as they appear in Table 1.

  The TDR/FDR results of Figure 11 show very insignificant variations with respect to the network size. This is somewhat expected since in wireless networks it is the routing that is mostly susceptible to scalability which is not a concern herein.

Below we give a summary of the test results of the proposed algorithm.

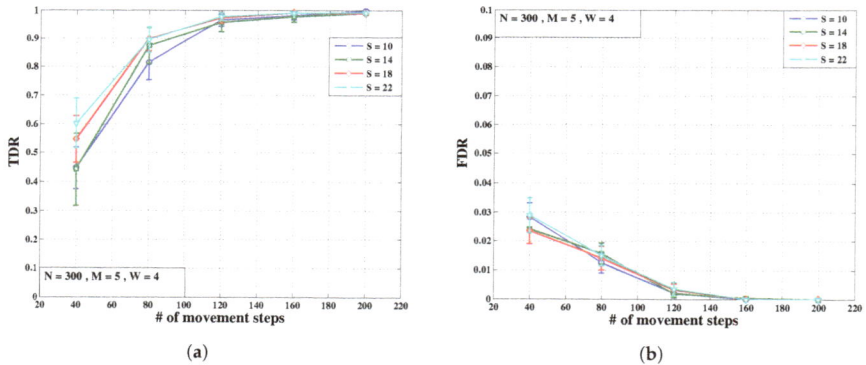

Figure 7. Variation of TDR (**a**) and FDR (**b**) versus time for different number of Sybil IDs.

(a)   (b)

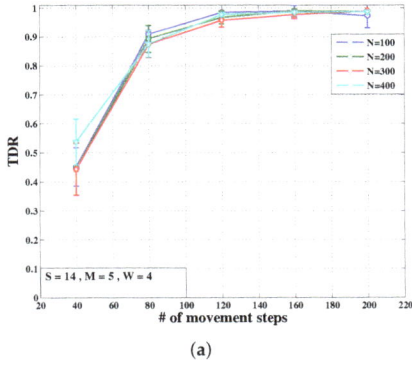

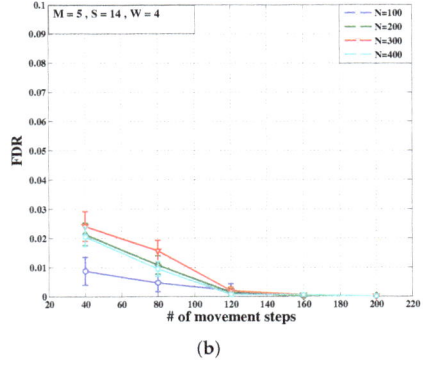

Figure 8. Variation of TDR (**a**) and FDR (**b**) versus time for different normal node populations.

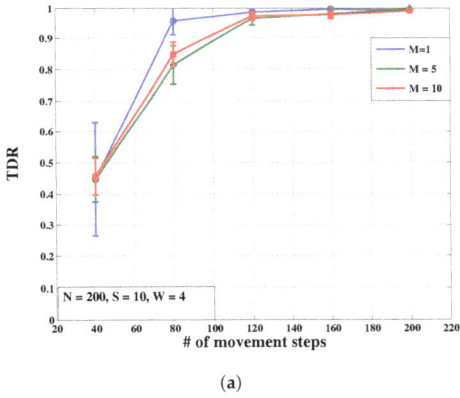

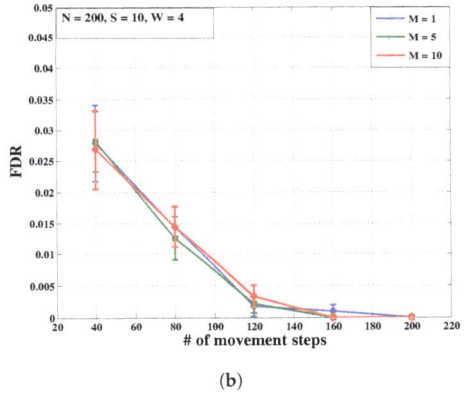

(a)   (b)

Figure 9. Time variations of TDR (**a**) and FDR (**b**) for different malicious node populations.

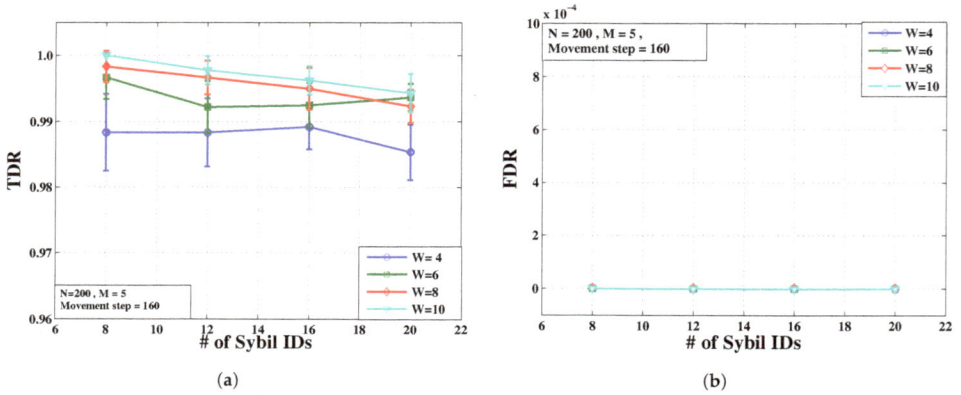

Figure 10. The effect of WD population on the variation of TDR (**a**) and FDR (**b**) versus the number of Sybil IDs.

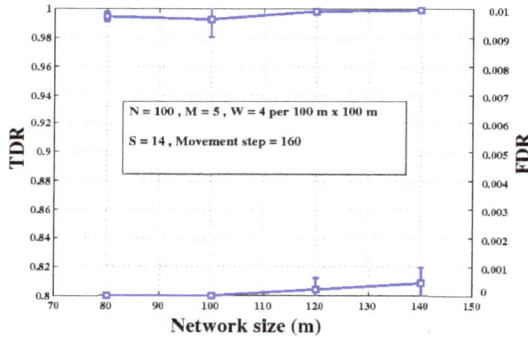

Figure 11. Scalability evaluation of the proposed algorithm.

- Observing Figures 7 and 9, TDRs and FDRs surely reach their steady-state values (upon sufficient information collection) and interestingly almost at the same time (at movement step greater than or equal to 120) regardless of the values of the network parameters.

- Comparing Figures 7 and 8, the convergence rates of TDR and FDR are more sensitive to the number of Sybil IDs per malicious node and the number of normal nodes respectively, before reaching their steady states. This is expected because of the very definition of TDR and FDR in Section 5.1.

- Somewhat related to the previous observation, Figures 7a and 9a show that increasing the total number of Sybil IDs by increasing S and M have opposite impacts on TDR. In fact, increasing S makes the Sybil pattern more noticeable and in favor of TDR. As illustrated by Figures 7b and 9b increasing the total number of Sybil IDs does not have much effect on FDR.

- The results in Figure 10a suggest that, in steady state, there is a turning point of WD population above which not much TDR gain is achieved.

Regarding the three remaining measures of performance, in the proposed algorithm, the memory overhead is equal to the amount of memory needed for storing upper-triangle matrices A^k_{co-prs}

and C_{co-prs}^k which is equal to $O(H^2)$ (refer to Equation (1)). It should be noted that this memory consumption occurs only in WDs.

To do their jobs, WDs overhear the Hello Packets which are already in use in mobile sensor networks. After forming and updating C_{co-prs}^k, each WD sends its final C_{co-prs}^k to the designated WD just once, in the detection phase. So the communicational overhead is $O(H^2)$. It is very important to note that in calculating the communicational overhead, we do not consider Hello-Packets related overhead (and its consumed energy) since Hello-Packets broadcasting mechanism is embedded in any mobile wireless routing and is not additionally imposed by our proposed detection algorithm.

The algorithm includes arithmetic and comparison operations on upper-triangle matrices (with $H \times H$ dimensions) in WDs and the designated WD leading to a computational complexity of $O(H^2)$.

5.3.2. Comparison

In this section, we compare our algorithm with two methods in [24,25] in terms of TDR and FDR. These works are chosen due to the similarities in the adopted attack models and approach.

- **Comparison 1:** To compare our algorithm with PASID, we have to set the simulation parameters as per [24]. Therefore, in this part, we assume $M = 1$, *max speed* $= 0.2$ m/s, *wireless radio range* $= 10$ m, and the results are calculated at *movement step* $= 200$. Figure 12 illustrates TDR and FDR for PASID and our algorithm against the network size (herein called topography size as in [24]). To be consistent with results in [24], each point of the new proposed algorithm's results is an average over 240 combinations of $N = 5, 10, 25, 40$, and $S = 5, 10, 20$, and $W = \lfloor N/2 \rfloor$ where the latter corresponds to the best case result reported in [24]. The results of PASID are extracted from [24]. The new algorithm performs better than PASID in terms of FDR for all network sizes. In terms of TDR, the new algorithm has a significant edge over PASID in large networks while slightly under-performing it in smaller ones.
- **Comparison 2:** In this test we compare our algorithm with another relevant algorithm in [25]. Our algorithm uses a somewhat similar information collection strategy as in [25] while using a completely different detection rule. To make our results comparable, we adopt the same values for our simulation parameters as those in [25].

Figure 13 compares TDR and FDR variations versus time for different number of Sybil IDs. It should be noted that [25] provides the results just for a limited number of movement steps (time period) and no confidence interval consideration. Figure 13a suggests that initially TDR and its improvement rate in [25] is better than our algorithm. However, at some point our algorithm not only catches up but also starts out-performing [25]. Both algorithms expectedly show slightly better performance by increasing the number of Sybil IDs per malicious node (S). Regarding FDR, our algorithm starts significantly better and the difference in performance vanishes through time. Consistent with its definition, FDR in Figure 13b shows no sensitivity to variation of S. The FDR of our algorithm is initially much better than the FDR of the algorithm in [25]. However, both algorithms perform asymptotically perfect in the long term. Time variations of TDR and FDR with varying malicious node population for both algorithms are illustrated in Figure 14. We observe almost the same trends as in Figure 13 with the following particularities. TDR in [25] seems to decline after some time as apposed to the new algorithm wherein TDR keeps improving constantly. Moreover, only TDR in the new algorithm and only FDR in the algorithm of [25] show sensitivity to the malicious node population in the transient state. To sum up:

- Regarding TDR, although the detection process is slightly late in picking up due to its conservative approach (Equation (2) in step V) to somewhat compensate for erroneous receptions , it performs reliably and asymptotically better.
- Introducing the trap state (01/10) in the state diagram of Figure 13 and applying the detection rule of Equation (4) in step V result in an adequate FDR even in short term.

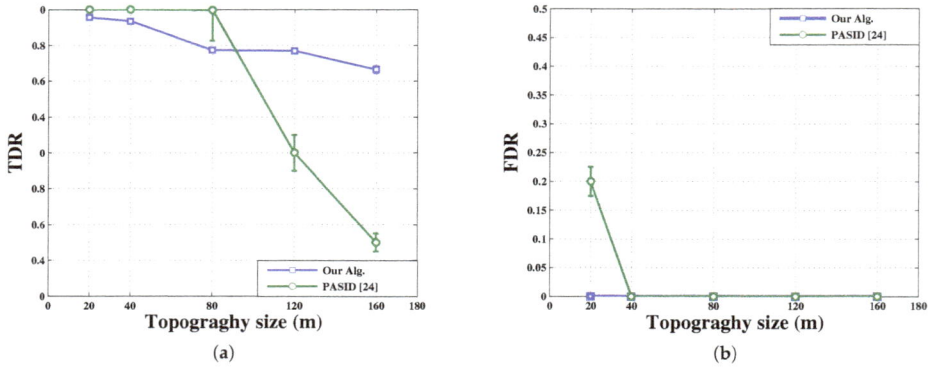

Figure 12. Performance comparison of PASID and our new proposed method in terms of TDR (**a**) and FDR (**b**) versus the topography.

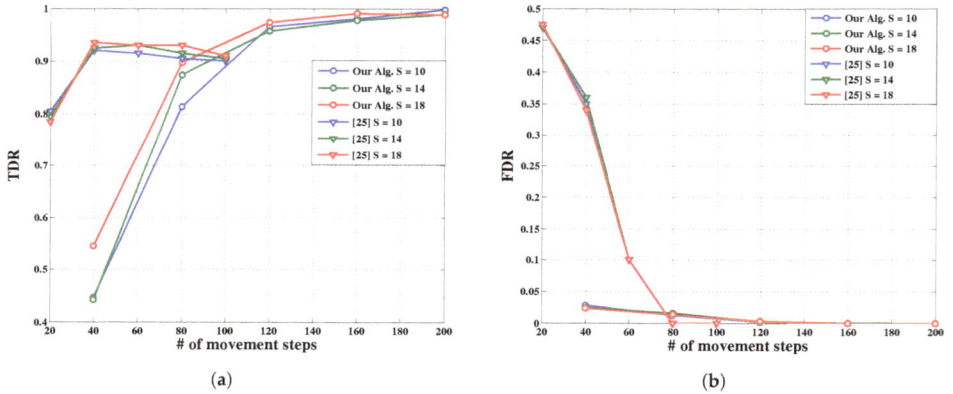

Figure 13. Performance comparison of [25] and our new proposed method in terms of TDR (**a**) and FDR (**b**) versus number of movement steps for varying number of Sybil IDs.

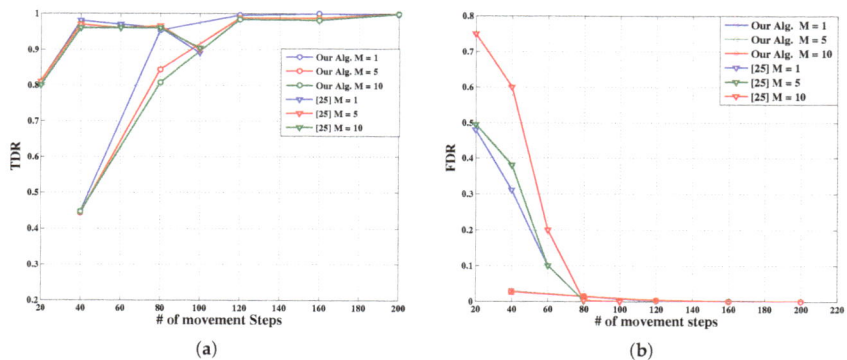

Figure 14. Performance comparison of [25] and our new proposed method in terms of TDR (**a**) and FDR (**b**) versus number of movement steps for varying number of malicious nodes.

Finally, we would like to point out that the diagrams representing variations of TDR and FDR versus the number of normal nodes in Figure 10 of [25] are not usable for comparison purposes.

6. Conclusions

In this paper, a new computationally lightweight method for detecting Sybil IDs in mobile WSNs is proposed. The proposed algorithm employs watchdog nodes which overhear Hello Packets exchanges between nodes. Each watchdog node uses a newly introduced co-presence state diagram to produce partial detection information. A designated watchdog node aggregates all partial detection information and uses a new detection rule to generate the final Sybils list. The new algorithm features low extra communication overhead, as well as a satisfactory compromise between two otherwise contradictory detection measures of performance, TDR and FDR.

The performance of the new algorithm is evaluated and compared with other similar important recent watchdog-based algorithms using extensive simulations. The simulation results illustrate the merits of the algorithm collectively.

Author Contributions: The authors contributed equally to this work.

Conflicts of Interest: The authors declare no conflict of interest.

Abbreviations

The following abbreviations are used in this manuscript:

WSN	Wireless Sensor Network
TDR	True Detection Rate
FDR	False Detection Rate
PHY	Physical
RSSI	Received Signal Strength Indicator
TDoA	Time Difference of Arrival
AoA	Angle of Arrival
IoT	Internet of Things
MANET	Mobile Ad hoc Network
PASID	Passive Ad hoc Sybil Identity Detection
UWB	Ultra-Wide Band
USAS	Unpredictable Software-based Solution
WD	Watchdog Node
SN	Sensor Node
MN	Malicious Node

References

1. Akyildiz, I.F.; Su, W.; Sankarasubramaniam, Y.; Cayircl, E. A survey on sensor networks. *IEEE Commun. Mag.* **2002**, *40*, 102–114.
2. Karlof, C.; Wagner, D. Secure routing in wireless sensor networks: Attacks and countermeasures. *Ad Hoc Netw.* **2003**, *1*, 293–315.
3. Sagduyu, Y.E.X.; Berry, R.A.; Ephremides, A. Jamming Games in Wireless Networks with Incomplete Information. *IEEE Commun. Mag.* **2011**, *49*, 112–118.
4. Tsiropoulou, E.E.; Baras, J.S.; Papavassiliou, S.; Qu, G. On the Mitigation of Interference Imposed by Intruders in Passive RFID Networks. In Proceedings of the 7th International Conference GameSec, New York, NY, USA, 2–4 November 2016.
5. Chen, X.; Makki, K.; Yen, K.; Niki, P. Sensor Network Security: A Survey. *IEEE Commun. Surv. Tutor.* **2009**, *11*, 5–73 .
6. Newsome, J.; Shi, E.; Song, D.; Perrig, A. The Sybil Attack in Sensor Networks: Analysis & Defenses. In Proceedings of the Third International Symposium on Information Processing in Sensor Networks, Berkeley, CA, USA, 26–27 April 2004.

7.	Bettstetter, C.; Resta, G.; Santi, P. The Node Distribution of the Random Waypoint Mobility Model for Wireless Ad Hoc Networks. *IEEE Trans. Mob. Comput.* **2003**, *2*, 257–269.

8.	Ssu, K.F.; Wang, W.T.; Chang, W.C. Detecting Sybil attacks in wireless sensor networks using neighboring information. *Comput Netw.* **2009**, *53*, 3042–3056.

9.	Seshadri, A.; Perrig, A.; Doorn, L.; Khosla, P. SWAtt: Software-based attestation for embedded devices. In Proceedings of the IEEE Symposium on Security and Privacy, Berkeley, CA, USA, 12 May 2004.

10.	Seshadri, A.; Perrig, A.; Doorn, V.; Khosla, P. SCUBA: Secure Code Update By Attestation in Sensor Networks. In Proceedings of the 5th ACM Workshop on Wireless Security, Los Angeles, CA, USA, 29 September 2006.

11.	Jin, X.; Putthapipat, P.; Pan, D.; Pissinou, N.; Kami Makki, S. Unpredictable Software-based Attestation Solution for Node Compromise Detection in Mobile WSN. In Proceedings of the Workshop on Advances in Communications and Networks, Miami, FL, USA, 6–10 December 2010.

12.	Borisov, N. Computational puzzles as Sybil defense. In Proceedings of the IEEE international Conference on Peer-to-Peer Computing (P2P), Cambridge, UK, 6–8 September 2006.

13.	Zhang, Q.; Wang, P.; Reeves, D.; Ning, P. Defending against Sybil attacks in sensor networks. In Proceedings of the 25th IEEE International Conference Distributed Computing Systems Workshops, Columbus, OH, USA, 6–10 June 2005.

14.	Sharmila, S.; Umamaheswari, G. Node ID based detection of Sybil attack in mobile wireless sensor network. *Int. J. Electron.* **2012**, *100*, 1441–1454.

15.	Marti, S.; Giuli, T.J.; Lai, K.; Baker, M. migration watchdog Mitigating Routing Misbehavior in Mobile Ad Hoc Networks. In Proceedings of the 6th Annual International Conference on Mobile Computing and Networking, Boston, MA, USA, 6–11 August 2000.

16.	Sun, B.; Osborne, L.; Yang, X.; Guizani, S. Intrusion detection techniques in mobile ad hoc and wireless sensor networks. *IEEE Wirel. Commun.* **2013**, *14*, 56–63.

17.	Douceur, J.R.; Druschel, P.; Kaashoek, M.F.; Rowstron, A. The Sybil attack. In Proceedings of the International Workshop on Peer-to-Peer Systems, Cambridge, MA, USA, 7–8 March 2002.

18.	Abbas, S.; Merabti, M.; Llewellyn-Jones, D.; Kifayat, K. Lightweight Sybil Attack Detection in MANETs. *IEEE Syst. J.* **2013**, *7*, 236–248.

19.	Demirbas, M.; Song, Y. An RSSI-based scheme for Sybil attack detection in wireless networks. In Proceedings of the International Symposium on World of Wireless Mobile and Multimedia Networks, Washington, DC, USA, 26 June 2006.

20.	Wen, M.; Li, H.; Zheng, Y.-F. TDOA-based Sybil attack detection scheme for wireless sensor. *J. Shanghai Univ.* **2008**, *12*, 66–70.

21.	Sarigiannidis, P.; Karapistoli, E.; Economides, A.A. Detecting Sybil Attacks in Wireless Sensor Networks using UWB Ranging-based Information. *Expert Syst. Appl.* **2015**, *42*, 7560–7572.

22.	Li, F.; Mitial, P.; Cocsar, M.; Borisov, N. SybilControl: Practical Sybil defense with computational puzzles. In Proceedings of the 19th ACM Conference on Computer and Communications Security, Raleigh, NC, USA, 16–18 October 2012.

23.	Tangpong, A.; Kesidis, G.; Hsu, H.; Hurson, A. Robust Sybil Detection for MANETs. In Proceedings of the International Conference on Computer Communications and Networks (ICCCN), San Francisco, CA, USA, 3–6 August 2009.

24.	Piro, C.; Shields, C.; Neil Levine, B. Detecting the Sybil Attack in Mobile Ad hoc Networks. In Proceedings of the Secure-comm and Workshops, Baltimore, MD, USA, 28 August–1 September 2006.

25.	Jamshidi, M.; Zangeneh, E.; Esnaashari, M.; Meybodi, M. A lightweight algorithm for detecting mobile Sybil nodes in mobile wireless sensor networks. *Comput. Electr. Eng.* **2016**, *40*, 1–13.

26.	Sobeih, A.; Hou, J.C.; Kung, L.; Ning, L.; Zhang, H.; Chen, W.; Tyan, H.; Lim, W. J-Sim: A simulation and emulation environment for wireless sensor networks. *IEEE Wirel. Commun.* **2006**, *13*, 104–119.

future internet

MDPI

Article

An Adaptive Privacy Protection Method for Smart Home Environments Using Supervised Learning

Jingsha He [1,2,*], Qi Xiao [1], Peng He [2,*] and Muhammad Salman Pathan [1]

[1] Faculty of Information Technology & Beijing Engineering Research Center for IoT Software and Systems, Beijing University of Technology, Beijing 100124, China; xqnssa@163.com (Q.X.); salman_htbk@hotmail.com (M.S.P.)

[2] College of Computer and Information Technology, China Three Gorges University, Yichang 443002, China

[*] Correspondence: jhe@bjut.edu.cn (J.H.); hpeng@ctgu.edu.cn (P.H.); Tel.: +86-135-0102-9135 (J.H.); +86-139-7200-2550 (P.H.)

Academic Editor: Georgios Kambourakis
Received: 23 December 2016; Accepted: 1 March 2017; Published: 5 March 2017

Abstract: In recent years, smart home technologies have started to be widely used, bringing a great deal of convenience to people's daily lives. At the same time, privacy issues have become particularly prominent. Traditional encryption methods can no longer meet the needs of privacy protection in smart home applications, since attacks can be launched even without the need for access to the cipher. Rather, attacks can be successfully realized through analyzing the frequency of radio signals, as well as the timestamp series, so that the daily activities of the residents in the smart home can be learnt. Such types of attacks can achieve a very high success rate, making them a great threat to users' privacy. In this paper, we propose an adaptive method based on sample data analysis and supervised learning (SDASL), to hide the patterns of daily routines of residents that would adapt to dynamically changing network loads. Compared to some existing solutions, our proposed method exhibits advantages such as low energy consumption, low latency, strong adaptability, and effective privacy protection.

Keywords: smart home; privacy; FATS attack; supervised learning

1. Introduction

The most profound technologies are those that eventually disappear. They weave themselves into the fabric of every day life, until they become indistinguishable. Wireless sensors are becoming ubiquitous in smart home applications and residential environments. Smart home applications integrate multiple Internet of Things (IoT) devices and services that store, process, and exchange data. Micro-controllers can be used to analyze the status of the sensors for identifying events or the activities of residents. They then respond to these events and activities by controlling certain mechanisms that are built within the home. A simple example of such a smart behavior is to turn on the lights when a person enters a room [1]. This can be realized by triggering an infrared sensor when the person enters the room, and the micro-controller can combine the activity and the brightness of the room, to determine whether the lights should be turned on.

There will be a multitude of devices in a wireless sensor network (WSN) around residential environments. These monitoring devices can be classified into three categories: sensors, physiological devices, and multimedia devices. Sensors are used to measure the environmental parameters. Physiological devices monitor health conditions and vital signs. Multimedia devices capture audiovisual information and provide an interface between the system and the user [2].

A sensor in a residence is a simple autonomous host device that can sense a phenomenon, convert signals into data, process the data, and then transmit it to a sink node for further analysis [3]. However, the societal concerns of smart home technology evolution, in relation to the privacy and security of the citizen, appear to be at an embryonic stage [4]. Although smart home technologies can bring a great deal of convenience to residents, it is also possible that people's daily behaviors in such an environment would become exposed to attackers, who can use the smart home in a malicious way. Therefore, the issue of privacy protection in the smart home environment has become one of the most important challenges.

In the smart home scenario, almost all of the sensor nodes only transmit information when a related event is detected, which is called event-triggered transmission. A global adversary has the ability to monitor the traffic of the entire sensor network, and thus, can immediately detect the origin and time of event-triggered transmissions. Although encryption algorithms can be used to protect data in the transmission, the emergence of new types of attack methods can make such traditional approaches invalid. Such attacks only need access to the timestamp and fingerprint data of each radio message, and in a wireless environment, the fingerprint is a set of features of a radio frequency waveform that are unique to a particular transmitter. Thus, the primary attacks that we are concerned with here are the Fingerprint And Timing-based Snooping (FATS) attacks [5], which have been shown to be very effective in inferring the Activity of Daily Livings (ADLs) of the residents.

The most simple and effective way of resisting FATS attacks is to inject fake messages into the transmission sequence. There have been extensive studies on the protection of the privacy of residents in a smart home environment, by taking into consideration the limitations of communication bandwidth, battery energy, and computing power. Most solutions proposed so far are based on a fixed frequency or probabilistic model, which make it hard to identify the real messages in the sequence of messages, even if the attacker can access the global information. These solutions have a major drawback, however, i.e., the reporting of a real event could be delayed until the next scheduled transmission. The delay of sensed data can cause degradation of the Quality of Service (QoS) in many applications, especially in those with intelligent sensing, where sensor data need to be obtained in real time in order to make decisions. To address this problem of delay, Park et al. proposed a method based on behavioral semantics [6]. However, this method would depend on the accuracy of the prediction, meaning that, if the prediction of the next message was not accurate, the added fake messages would not be enough to disturb the statistical analysis, and the ADLs of the residents would still be exposed. In this paper, we propose a method to protect against FATS attacks. The method is based on sample data analysis and supervised learning, which can adapt to network loads, as well as to the common living habits derived from the real data.

The remainder of this paper is organized as follows. In Section 2, we describe the FATS attack model and introduce some existing solutions. In Section 3, we make some assumptions about the network environment and the adversary, and describe the requirements of the privacy protection method. In Section 4, we describe our method in detail. In Section 5, we compare our method to some existing methods, to demonstrate the advantages of our method. Finally, in Section 6, we conclude this paper and also describe our future work. At the end of the paper, we list all of the acronyms used throughout this paper as the appendix.

2. Related Work

In this section, we will first introduce the FATS attack model and then briefly describe some of the existing solutions for resisting FATS attacks. We will also analyze the deficiencies of the existing solutions.

2.1. The FATS Attack Model

The FATS attack model focuses on collecting the fingerprints and timestamps to gain access to the behavior of the residents, even if the sensor data is protected by using a sufficiently secure and reliable encryption method. The attack model is shown in Figure 1 and the four tiers of attacks are described below:

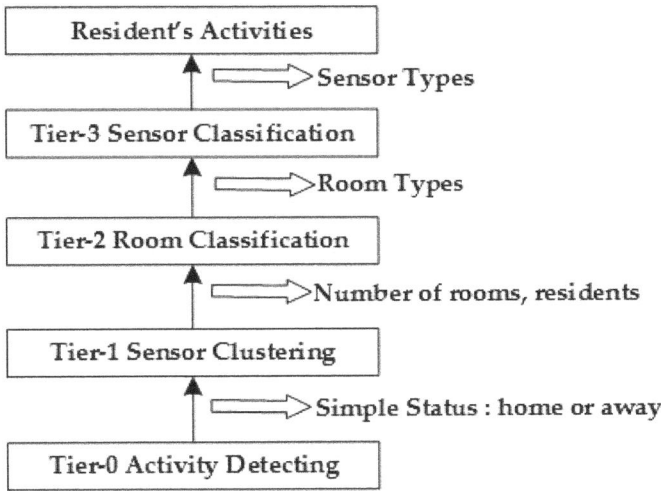

Figure 1. Four tiers in the FATS (Fingerprint And Timing-based Snooping) attack model.

Tier-0: General Activity Detection. The adversary can only detect very general activities, such as home occupancy or sleeping.

Tier-1: Sensor Clustering. It assumes that a particular sensor that is triggered during a timestamp will be very close in space to infer other sensors in the same room. In this tier, the number of rooms and people in the home can be predicted.

Tier-2: Room Classification. The main goal in this tier is to identify the features of the rooms via an analysis of the previous cluster in Tier-1. At this tier, the attacker can ascertain the layout of the house and can predict the residents in the rooms. The privacy of the residents can be infiltrated by the attacker.

Tier-3: Sensor Classification. In this tier, the goal is to identify the activities in the home, such as cooking, showering, and so on. The attacker can calculate a feature vector for each sensor from the answers in Tier-2 and can then classify each sensor by using Linear Discriminant Analysis (LDA).

Through analyzing the above tiers, residents' behavior can be exposed. First, a feature vector of every temporal activity cluster in every device can be calculated. Then, by importing the vector to the LDA classifier that was trained for other homes, the hand-labeled activity labels can be used to distinguish real activities. This approach can recognize many daily activities, including showering, washing, grooming, cooking, etc. [4].

2.2. Methods to Resist the FATS Attack

In the ConstRate (sending messages in constant rate) model, all of the sensor nodes in the network maintain the same frequency when sending messages, whether or not actual events happen. When a real event occurs, it has to wait until the next transmission. Therefore, the model can effectively resist the static analysis by the attacker with a global listening ability. This model also has a congenital deficiency, i.e., the average delay is half of the transmission interval and it is difficult to determine the transmission interval. If the transmission rate is low, the delay will be very high; whereas, if the rate is high, the delay will decline, but the number of fake messages will increase significantly, resulting in an increased energy consumption.

Yang et al. proposed a probability-based model called the FitProbRate model, that aims at reducing the latency of a fixed frequency transmission [7]. The main idea is that every sensor in the network sends messages with an interval that follows the exponential distribution. When a node detects a real event, the algorithm needs to identify a minimum interval to obey the exponential distribution, and then waits to send the real event. Following this, all adjacent intervals will follow the same distribution, making the attacker unable to determine the real messages from the transmission sequence. This model can reduce the delay of transmission in some situations, e.g., when the time interval is relatively average and the sending interval is slightly longer. If the time interval is not uniform and is frequently triggered with small intervals, the delay will become high.

Park et al. proposed a behavioral semantics model to generate a small amount of fake data, to protect the activity that will happen in the near future [7]. Firstly, the model predicts the activity from a long-term history and then presents the sensor nodes with the forecast of the activity sequence. If an attacker listens to a sequence in order to monitor the sensors, the attacker can only predict misbehavior. This model adds fake messages to disturb the FATS attack in Tier-3 and privacy protection depends on the accuracy of the prediction of future behaviors. Therefore, if the prediction is not accurate, the added fake messages will not make much difference. Generally, the reliability of this scheme is lower than the ConstRate and FitProbRate models.

2.3. Summary of the Related Work

2.3.1. The intervals of the Send Sequence

In the ConstRate and FitProbRate models, the interval between the fake messages and the real messages is subject to the same distribution. The purpose of adding fake messages is to prevent the attacker from distinguishing between real and fake messages. If an attacker is not able to recognize the real messages from the message sequence sent by the sensor node, the purpose of adding noise is successfully achieved. Consequently, assuming that an adversary monitors the network over multiple time intervals, in which some intervals contain real event transmissions and some do not, if the adversary is unable to distinguish between the intervals with significant confidence, the real event will be hidden [8]. If the sensor nodes have a sufficient randomness to send fake and real messages, it makes it less likely that the adversary will be able to recognize the fake data from the transmission sequence, resulting in the analysis of the wrong ADL, and thus, the protection of the privacy of the residents.

2.3.2. The Traffic of the Whole Network

To reduce the delay of data transmission, the transmission interval should be reduced, resulting in a high traffic load, as well as a significant increase in the probability of collision. As Figure 2 shows, in an actual sensor network, the closer it gets to the sink node, the larger the amount of data that needs to be forwarded. If all of the sensor nodes send fake messages following the same model, the nodes that are closest to the sink node will have to assume a load which is too heavy to forward messages. In our method, a sensor node will adapt to its own network status when sending fake messages, i.e., a node will add a lower number of fake messages when the forwarding load appears to be heavy. Such an approach would result in the three types of nodes consuming energy at a more balanced level, prolonging the life of the network in comparison to other models.

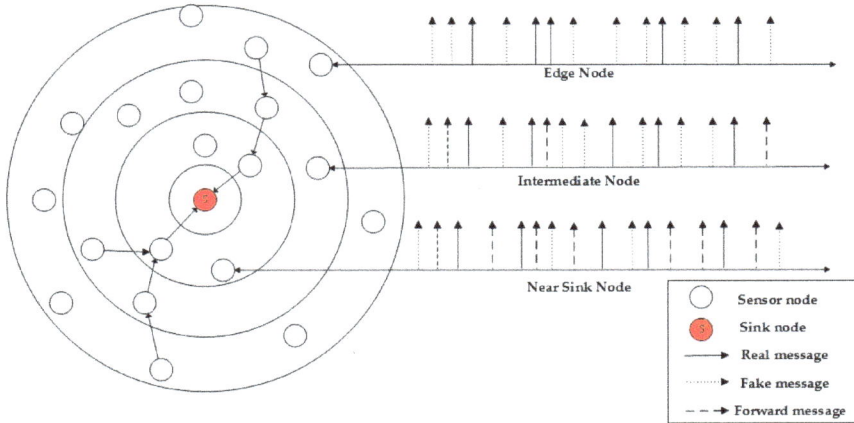

Figure 2. Traffic status of the sensors.

2.3.3. The Particularity of Smart Home Environment

In the smart home environment, when people go to sleep or go out to work, there are very few real events triggered during the corresponding time slots. It seems that there is no need to add fake messages during such periods, due to few regular activities in the smart home. Accordingly, the ideal situation should be that the sensor nodes only send fake messages when the residents are present, with some daily activities taking place. Thus, the number of fake messages can be reduced. However, the adversary can learn the routines of individual residents through analyzing the sending patterns of the sensor nodes. Following the ideas of k-anonymity, we will thoroughly analyze the data to obtain the routines of the public, so that message sending will be carried out in such a way that it would be very difficult for the attacker to detect the routines of individual residents.

3. Assumptions and Requirements

In this section, we describe the network model and some assumptions about the adversary in SDASL, as well as the requirements of privacy protection that will guide the design of our privacy protection method.

3.1. The Network Model

Similar to other WSNs [9], nodes in smart home consist of the sink node (only one sink node in the sensor network) and sensor nodes $N = \{N_1, N_2, N_3, N_4 \ldots N_n\}$. As shown in Figure 3, the smart home provider is a reliable service provider who can collect and analyze data from many homes. The sink node has a high enough computing power to take on complicated operations. The sensor node is the smallest unit in the WSN, and has a limited computing power and limited battery capacity. Sensor nodes can apply encryption algorithms to encrypt collected data [10], before sending it to the sink node. We assume that the encryption algorithms used are safe enough and that the attacker cannot acquire the original data through analyzing the cipher texts. We also assume that the sink node is actively powered, while being equipped with tamper-resistant hardware [11]. Consequently, it is reasonable to assume that the adversary cannot compromise the sink node.

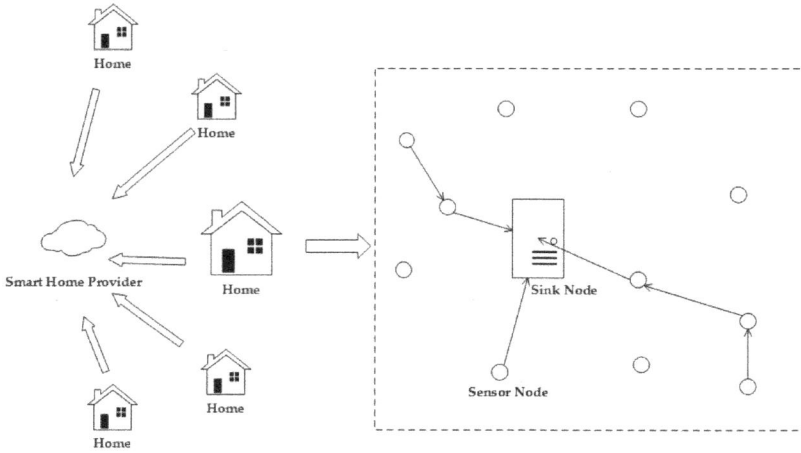

Figure 3. The framework of a smart home.

We assume that the dataset collected at the sink node is $D = \{X, y\}$ and $X = \{X_1, X_2, X_3 \cdots, X_n\}$, where X_i refers to the set of sensor data from sensor node N_i. A datum of X_i is denoted as $X_i^{(j)} = \{x_1, x_2, x_3, x_4\}$, in which x_1 is the sensor data, x_2 is the traffic status, x_3 is the timestamp of transmitting the sensor data, and x_4 is the flag to indicate whether the message is real or fake. Label y in D can only be a binary value $\{0,1\}$. The minimum unit of the collected datasets is $\left\{ X_j^{(i)}, y^{(i)} \right\}$, which means that the ith sensor data comes from sensor node N_j with label $y^{(i)}$.

3.2. The Adversary Model

The global adversarial model used in this paper is similar to the one that is considered as external, passive, and global [12,13]. In contrast to the passive adversary, an external adversary cannot control any nodes in the WSN. Instead, it can only monitor, eavesdrop, and analyze the communication in the network, via channel snooping. After obtaining the transmission status, the global adversary can identify the behavior of people in the smart home by applying static analysis, using a method such as the FATS Model.

3.3. Requirements of Privacy Protection

The particularity and sensitivity of the smart home makes it important to consider the privacy, energy efficiency, and latency of the sensor network in the design of privacy protection methods.

3.3.1. Privacy

The WSN in the smart home environment collects sensitive data about people living in the environment. A good privacy protection model should be robust to resist such attacks as FATS, as well as statistical analysis that can acquire private activities. The main purpose is that, even if an attacker can listen to a global message sequence along with the time of transmission, it is still not possible for the attacker to identify fake messages from the transmission sequence, i.e., the added fake messages would prevent the attacker from obtaining the desired results.

3.3.2. Energy Efficiency

It is not considered a good privacy protection scheme if the implementation of the scheme would reduce the lifetime of the entire WSN. A good protection scheme should keep the overhead as low as possible, to extend the life of the WSN as much as possible. However, adding fake messages can incur extra energy consumption. In our method, we take the average load of the traffic into consideration. When the traffic load is high, the algorithm reduces the number and frequency of the fake messages accordingly. Packet loss and collision are also reduced as a result.

3.3.3. Low Latency

As the aging of the population becomes a serious issue, the smart home would make a great contribution to improving the quality of life of the elderly [14]. In such applications, it is necessary to trigger events in a timely manner, to immediately send the sensed data to the sink node, to determine or predict the abnormal situation (especially physical health status) of the elderly, and to take appropriate actions. If the latency was too long, it would take a lot of time for the sensed data to be received, lowering the efficiency of the timely treatment of elderly people. A good privacy protection model should keep the latency within the normal range of acceptance, allowing the sink node to make timely decisions, in order to meet the requirements of the applications.

4. The Proposed Model

In this section, we introduce our adaptive method, which is based on supervised learning. The proposed method consists of three separate algorithms. Algorithm 1 is designed for sample data analysis. The smart home provider uses the algorithm to analyze the sample dataset generated in real smart home scenarios. Through the analysis, we can use the frequency distribution of the RF radio (FDR) to simulate the distribution of fake messages, ensuring that the frequency rate is similar to that in the sample datasets. Algorithm 2 is designed for supervised learning. Firstly, we analyze the load and time characteristics of the collected sensor data and label the results of Algorithm 1 accordingly. Then, we use supervised learning to generate the parameters of the prediction model. Finally, the sink node sends the parameters to all of the sensor nodes. Algorithm 3 is designed to allow the sensor nodes to update the parameters and to send fake messages. In the rest of this section, we will describe the three algorithms in detail.

4.1. Sample Data Analysis

In the smart home environment, the likely categories of privacy protection include scenarios of the user's going to work, coming back home, or going to sleep, etc. If fake data messages are only sent when the user is at home, to protect the user's behavior, then the behavior during the time at work, sleep, and other privacy-related periods, can still be leaked. We therefore design an algorithm that produces a transmission sequence which resembles that of a large amount of real data, to achieve better privacy protection. The adversary can only attain a general pattern of the people through monitoring the frequency of radio signals, thus protecting the privacy of individual residents. Consequently, we use FDR to describe the send frequency of the sensor network, which consists of one or more elements in the form [(start time, end time), average frequency].

Taking into account the different habits among people in different regions, the provider should include sample data from different regions in the analysis, and the procedure is shown in Figure 4. Firstly, we should count the number, as well as the time of message sending per minute, for which the answer is denoted as $F = \{(t_1, f_1), (t_2, f_2), (t_1, f_3) \ldots, (t_n, f_n)\}$. Then, we will use formula (1) to standardize f, for which the answer is denoted as $F^\wedge = \{(t_1, f_1'), (t_2, f_2'), \ldots, (t_n, f_n'))\}$. Secondly, we use the K-means clustering algorithm to cluster the elements in F', producing a group of periods over the 24-hour interval. Finally, we calculate the average frequency of sending in each of the periods and update the average frequency of sending, as well as the period p to the FDR. After the analysis, we will ascertain the FDR that will be sent to the smart home devices in every family.

$$f_i' = \frac{f_i - \min(F)}{\max(F) - \min(F)} \tag{1}$$

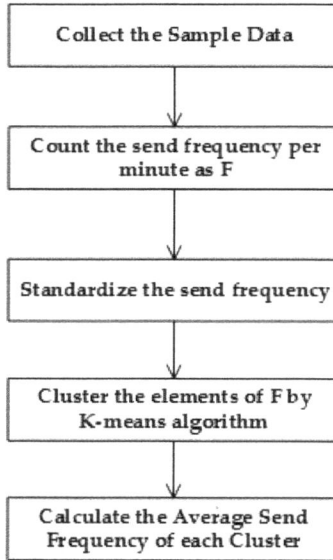

Figure 4. Analysis of sample datasets.

4.2. Supervised Learning

The supervised learning process is composed of three steps: collecting data, labeling collected data, and executing the learning algorithm and updating the parameters in the sensor nodes. All of the sensor nodes send fake messages in accordance with the initial time window, and the sink node labels the collected data in accordance with the FDR, as well as the traffic status. The learning algorithm will train the labeled data from different sensor nodes and the sink node will hand out the parameters to the sensor nodes.

4.2.1. Data Collection

Each sensor node works in accordance with Algorithm 1 of Figure 5 to generate information and decides whether or not to send fake messages. At the beginning of the algorithm, all of the sensor nodes in the WSN will send fake messages using a fixed time window. After training the learning algorithm, the prediction model will be plugged with a new θ, as shown in Figure 5. Each sensor node needs two variables to input into the prediction model, in order to determine whether or not to send fake data. The two variables are the traffic status and time, represented by x_1 and x_2, respectively. Traffic status is calculated using Formula (2), where ft represents the number of messages forwarded and st represents the number of messages sent within the time window.

$$x_1 = \frac{ft + st}{st}; \qquad x_2 = \frac{nowTimeStamp - dateStamp}{60 * 60 * 24} \qquad (2)$$

To normalize the time, we map the current timestamp in the range [0,24], which is represented using x_2. In x_2, the nowTimeStamp denotes the current timestamp and the dateStamp denotes the starting time of the day. These parameters are used as inputs for the prediction model. The prediction model consists of a series of operations, along with a hypothesis function. If the result of the prediction model is not smaller than 0.5, a fake massage should be sent after a random time delay. If the result is smaller than 0.5, the algorithm ascertain whether a fake message has been sent during the previous K iterations. If so, the time window will be doubled. Once real data is sent, the original time window will be resumed.

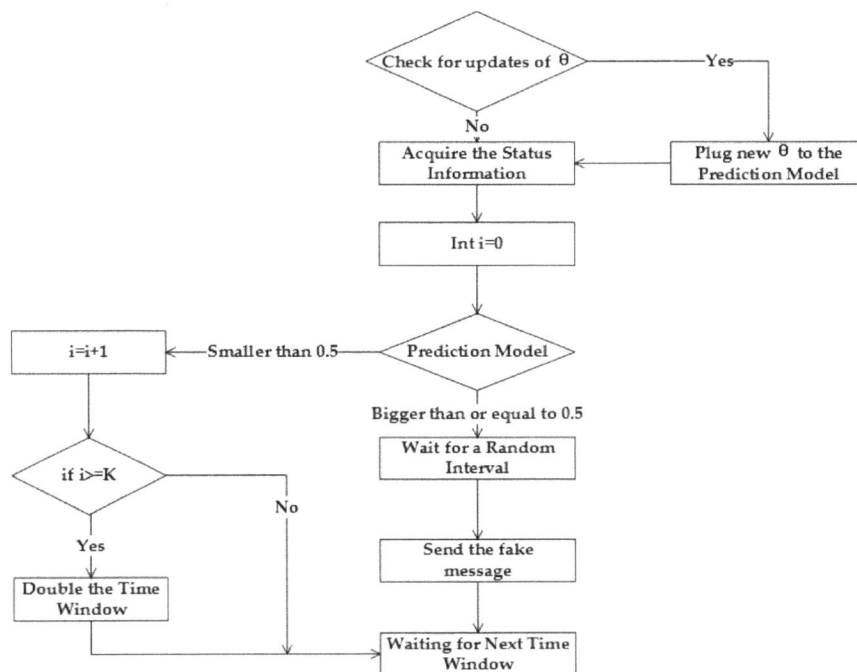

Figure 5. Sensor node determines whether or not to send fake messages.

4.2.2. Sensor Data Labeling

The sink node would classify received data into two categories: fake data and real data, which are marked using a flag. The main purpose of labeling is to mark the fake data with label 1, meaning that fake messages should be sent, and to use label 0 otherwise. The basis of the classification is the FDR and the traffic status. Firstly, we label fake messages with 0 during the period of sending messages sparsely, and with 1 during the period of sending messages at a higher rate. After labeling the fake messages, supervised learning will take place.

4.2.3. Learning and Parameter Updating

Supervised learning is one type of learning method, in which a model learns from the training data and then predicts new instances of an event. It results in unlabeled data to be labeled through previous experience and then applies the labeled data to the learning algorithm. After training, the parameters are generated to update the prediction model. Song et al. presented a solution for supervised learning [15] and our algorithm has been inspired by this solution.

Algorithm 2 is shown in Figure 6. The algorithm inputs the datasets of sensor data and the FDR. The learning algorithm will deal with all of the data and calculate θ for each sensor node, respectively.

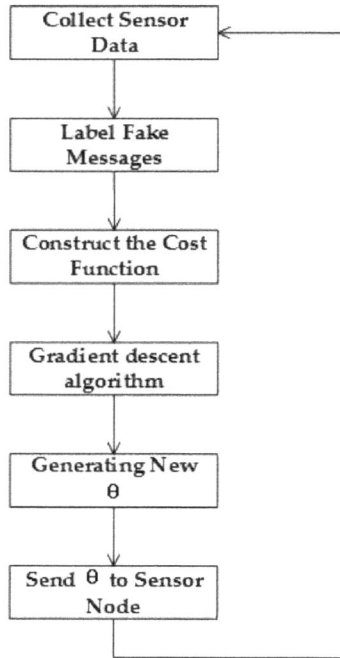

Figure 6. Supervised learning running in the sink node.

We use the logistic function as the hypothesis function, as expressed in Formula (3), where $\theta = [\theta_0, \theta_1, \theta_2 \cdots, \theta_w]^T$. In the formula, $h_\theta(x)$ is a $w+1$ dimensional parameter vector that we are learning from the training set, w is the number of features for the training, and θ_0 is the initial value for logistics regression. The learning algorithm and the hypothesis function can realize the goal of logistics regression.

$$h_\theta(x) = \frac{1}{1 + \exp\left(-\theta^T x\right)} \tag{3}$$

The main purpose of the learning algorithm is to find θ, to accurately classify the training data into two categories. When the cost function in Formula (4) reaches the global minimum, θ is the optimal solution of the learning algorithm. In the formula, m is the amount of training data. In the proposed SDASL method, every sensor node has its own training dataset, and thus, each sensor node has its own parameters.

$$J(\theta) = -\frac{1}{m} \sum_{i=1}^{m} [y^{(i)} \cdot log h_\theta(x^{(i)}) + (1 - y^{(i)}) \log(1 - h_\theta(x^{(i)}))] \tag{4}$$

The gradient descent algorithm is used here to obtain an appropriate θ for the hypothesis function which is described in Formula (5), in which α indicates the length of the gradient descent. Because $J(\theta)$ is a convex function, we can be sure of finding the local optimal value, which is also the global optimum.

$$\theta_i := \theta_i - \alpha \frac{\partial}{\partial \theta_i} J(\theta) \tag{5}$$

5. Evaluation

In this section, we first describe the method and the setup of the experiment, and then present the results of the experiment in terms of privacy protection, delay, and energy consumption, in comparison to the ConstRate and the FitProbRate models.

5.1. Experiment Setup

We used the public dataset related to accurate activity recognition in a home setting [16] in the experiment. We also downloaded several datasets from WSU CASAS (Center for Advanced Studies in Adaptive Systems) as the sample datasets, to calculate the FDR for supervised learning in Algorithm 2.

The experiment was carried out in a macOS environment and a PHP language was used to process the original datasets, that are in the formats of txt or dat. To use the datasets conveniently, we used PHP to convert the datasets into a uniform format and stored the results in MySQL. The details of the datasets are listed in Table 1, which includes the sensor number, trigger time, sensor status, and real event. For example, M17 was triggered at 12:49:52 and the event is wash_hands begin. M17 was triggered again at 12:50:42 and the event is wash_hands end.

Table 1. Examples of the datasets.

ID	Sensor Number	Trigger Time	Sensor Status	Real Event
1	M17	2008-02-27 12:49:52	ON	Wash_hands begin
2	M16	2008-02-27 12:49:54	ON	-
3	AD1-B	2008-02-27 12:50:01	0.302934	-
4	M17	2008-02-27 12:50:42	OFF	Wash_hands end
5	M19	2008-02-27 12:51:01	ON	Cook begin
...	...	...	...	...
6	T10	2008-02-27 13:07:14.	26.5 °C	-

In evaluating the effectiveness of privacy protection, we propose using ACA, the average clustering accuracy. This is because, at Tier-1, the FATS attack would try to cluster the sensor data, which plays a critical role in the attack. Should a sensor node be clustered into the wrong group, the predicted ADL would be inaccurate [17], resulting in an ACA which falls within the range [0, 1]. The closer ACA gets to 1, the closer the clustering results will be to equaling the number of rooms and the distribution of the sensor nodes. Conversely, when the clustering result is inconsistent with the actual sensor distribution in the rooms, the ADL will be adequately protected.

We also propose using FVR, which is the result of the total number of fake messages divided by the total number of real messages. Since it is hard to compare the three models under different conditions, we can use FVR to unify the main influencing factors. We compared the delay, energy consumption, and effectiveness of protection by the three models, through changing the FVR.

To compare the energy consumption between the three models, we compared the FVR under the condition that the same level of privacy protection is provided. Because all of the three models take a noise-based approach, the more noise in the WSN, the higher the energy consumption [18]. When considering the energy consumed by running the algorithm, it is generally recognized that energy consumption by wireless communication is much higher than the energy consumption by computation. It has been found that 3,000 instructions are needed for a sensor node to transmit 1 bit of data over a distance of 100 meters [19]. Therefore, we can ignore the energy consumption of the algorithm in our experiment.

In the evaluation of our model, the sample datasets were analyzed to derive the FDR and the time window. Then, through a number of iterations in supervised learning that would update θ, the transmission of real and fake data was collected by the sink node. Finally, we measured the FATS attack in Tier-1 to calculate the ACA.

5.2. Experiment Results

The results of the FVR are shown in Figure 7a. At the beginning, as the supervised learning is not performing, the FVR can achieve its maximum value, which is 25. As time goes by, the FVR gradually declines and eventually levels out at around 13. Figure 7b shows the results of ACA. As time goes by, the ACA decreases gradually, from 0.8 to 0.4, and then waves around 0.3. In the beginning, with the default parameters, the learning algorithm results in very poor privacy protection. Seven days later, the ACA arrives at a stably low level, indicating that good privacy protection has been achieved. As time goes by, the ACA gets lower, along with a continuous decrease in FVR thanks to the execution of the learning algorithm from the sink node and the sending algorithm from the sensor node.

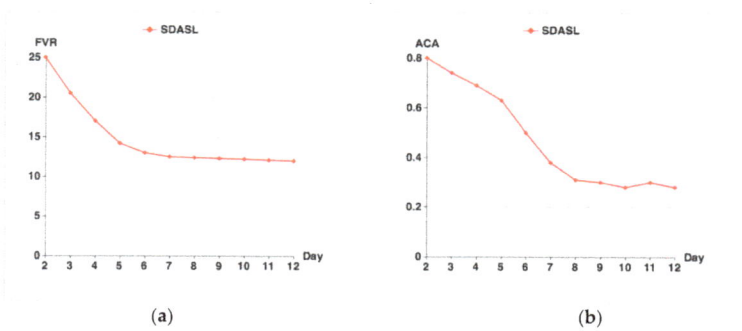

(a)   (b)

Figure 7. Relationship between FVR, ACA, and time in the SDASL model; (**a**) Description of the change of the FVR; (**b**) Description of the change of ACA.

Figure 8 shows that, with an increase in the FVR, the latencies of the ConstRate and the FitProbRate models gradually decrease. Since, in our method, real data is sent out without any delay, the latency is caused by the multi-hop forwarding time, which can be ignored, in contrast to the other two models. Figure 8 also provides the latency of a sensor node during a period of one day, which shows that the period from (0,6) to (9,17) has a lower latency due to a sparse transmission rate, and the period from (7,9) to (17,24) has a higher latency due to a dense transmission rate. This result indicates that the delay in the FitProbRate model is affected by the density of transmission, and as the transmission rate of the real event goes up, the delay will increase.

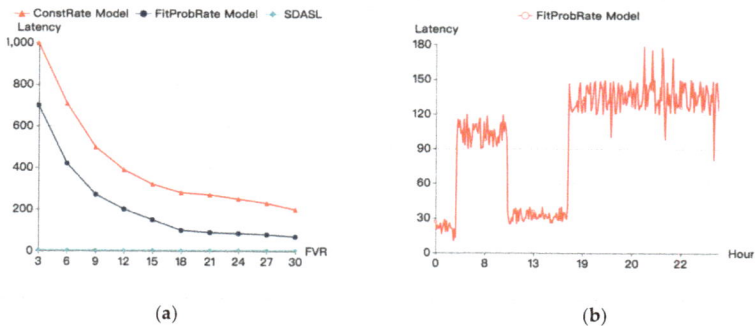

(a)   (b)

Figure 8. Comparison of latency; (**a**) Description of the change of the Latency which with the change of the FVR; (**b**) Description of the change of Latency of FitProbRate Model which with the change of the hours in a day.

Figure 9 shows that, with an increase in FVR, the ACA gradually decreases. While in the ConstRate model, the ACA remains at a low level, it is affected by FVR in our model and in the FitProbRate model. When FVR increases, ACA declines. As in the FitProbRate model, if real events happen frequently, the overall transmission will become dense. Whereas, if real events happen sparsely, the sequence will become synchronized. Therefore, the adversary can infer the routine of the residents by analyzing the density of the transmission sequence. Accordingly, when the sequence is dense, people may be at home and pursuing activities. When the sequence is sparse, people may be asleep or at work. In our model and in the ConstRate model, this problem does not occur.

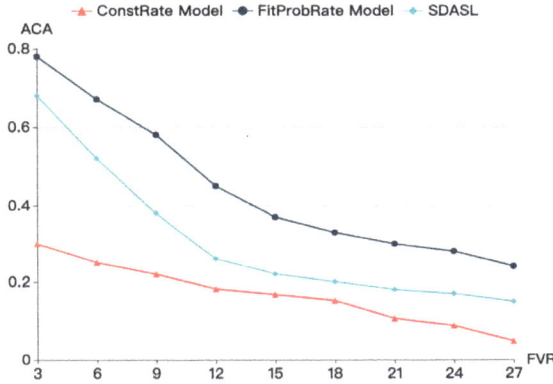

Figure 9. Comparison of ACA.

Figure 10 shows that with the increase in ACA, the FVR in all three models gradually decreases. Without considering the delay, the ConstRate model is the most energy-efficient model out of the three models. In the smart home, however, the delay must be controlled within an appropriate range. Compared to the FitProbRate and the SDASL models, for each ACA, the FVR in SDASL is much lower than that in FitProbRate, meaning that, under the same level of privacy protection, FitProbRate will incur 50% more noise data than SDASL. Since energy consumption mainly occurs during data transmission, SDASL will result in a 40% greater energy saving than FitProbRate, when the ACA is 0.2.

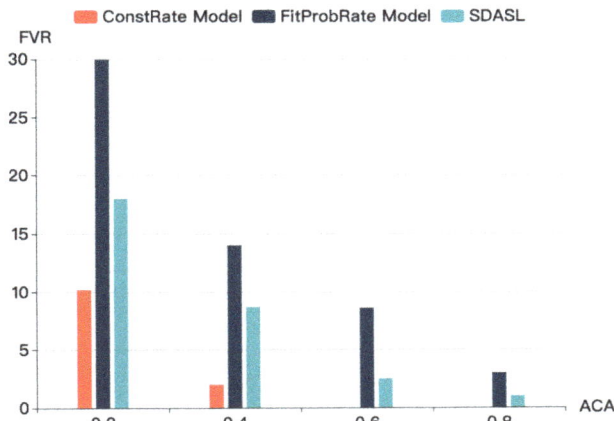

Figure 10. Comparison of energy consumption.

In summary, among the three models, ConstRate can provide the best privacy protection with an average latency and the longest delay. FitProbRate can reduce the delay, but a resident's routine may be revealed and the computation load in the sensor node would increase. Our SDASL model can overcome the shortcomings of the FitProbRate model, while reducing the latency.

6. Conclusions

In this paper, we proposed a new method to resist FATS attacks. Our method incorporates supervised learning to improve privacy protection, while analyzing sample data to provide the basis for data labeling. Compared to the ConstRate and the FitProbRate models, the experimental results clearly demonstrate the advantages of our SDASL model in terms of adaptiveness, low latency, and low power consumption, making it a better solution for smart home applications.

Following the approach in the proposed SDASL model, the collected data will be cached in the center of the smart home. When someone requests to access the data, the control module will decide whether or not to authorize the access. It may be true that the leakage of some insignificant information would not lead to the leakage of privacy. However, the attacker can still perform some analysis on such information, that could eventually lead to the leakage of privacy. In our future research, therefore, we will develop ways of resisting this type of attack, in order to improve privacy protection through access control.

Acknowledgments: The work in this paper has been supported by National High-tech R&D Program (863 Program) (2015AA017204).

Author Contributions: Jingsha He, ideas and general approach; Qi Xiao, model design, experiment and analysis; Peng He, general discussion and some contribution on experiment; Muhammad Salman Pathan, discussion and paper editing.

Conflicts of Interest: The authors declare no conflict of interest.

Appendix A

To help readers understand the acronyms clearly, we provide the following Table A1 to list all of the acronyms used throughout this paper.

Table A1. List of Acronyms and Explanations.

The Acronym	Explanation
ACA	Average Clustering Accuracy
ADLs	Activity of Daily Livings
ConstRate	Model that sends fake messages in constant rate
FDR	Frequency Distribution of RF radio
FATS	Fingerprint and Timing-based Snooping
FitProbRate	Model that is based on probability and looks for fit time interval to send fake messages
FVR	FVR is calculated by dividing the number of fake messages by the number of real messages
IoT	Internet of Things
QoS	Quality of Service
SDASL	Model that is based on sample data analysis and supervised learning
WSN	Wireless Sensor Network
WSU CASCA	Washington State University Center for Advanced Studies in Adaptive Systems

Future Internet **2017**, *9*, 7

References

1. De Silva, L.C.; Morikawa, C.; Petra, I.M. State of the art of smart homes. *Eng. Appl. Artif. Intell.* **2012**, *25*, 1313–1321. [CrossRef]
2. Alam, M.R.; Reaz, M.B.I.; Ali, M.A.M. A Review of Smart Homes-Past, Present, and Future. *IEEE Trans. Syst. Man Cybern. Part C Appl. Rev.* **2012**, *42*, 1190–1203. [CrossRef]
3. Ding, D.; Cooper, R.A.; Pasquina, P.F.; Fici-Pasquina, L. Sensor technology for smart homes. *Maturitas* **2011**, *69*, 131–136. [CrossRef] [PubMed]
4. Sanchez, I.; Satta, R.; Fovino, I.N.; Baldini, G. Privacy leakages in Smart Home wireless technologies. In Proceedings of the International Carnahan Conference on Security Technology, Rome, Italy, 1–6 May 2014.
5. Srinivasan, V.; Stankovic, J.; Whitehouse, K. Protecting your daily in-home activity information from a wireless snooping attack. In Proceedings of the 10th International Conference on Ubiquitous Computing, Seoul, Korea, 26 August 2008; pp. 202–211.
6. Park, H.; Basaran, C.; Park, T.; Sang, H.S. Energy-Efficient Privacy Protection for Smart Home Environments Using Behavioral Semantics. *Sensors* **2014**, *14*, 16235–16257. [CrossRef] [PubMed]
7. Yang, Y.; Shao, M.; Zhu, S.; Cao, G. Towards Statistically Strong Source Anonymity for Sensor Networks. *J. IEEE INFOCOM* **2008**, *9*, 51–55. [CrossRef]
8. Alomair, B.; Clark, A.; Cuellar, J.; Poovendran, R. Toward a Statistical Framework for Source Anonymity in Sensor Networks. *J. Mob. Comput.* **2011**, *12*, 1–6. [CrossRef]
9. Zhang, W.; Song, H.; Zhu, S.; Cao, G. Least privilege and privilege deprivation: Towards tolerating mobile sink compromises in wireless sensor networks. *Proc. ACM Mobihoc* **2005**, *4*, 2237–2258.
10. Klaoudatou, E.; Konstantinou, E.; Kambourakis, G. A Survey on Cluster-Based Group Key Agreement Protocols for WSNs. *Commun. Surv. Tutorials IEEE* **2011**, *13*, 429–442. [CrossRef]
11. Lin, X.; Lu, R.; Shen, X. MDPA: multidimensional privacy-preserving aggregation scheme for wireless sensor networks. *J. Wirel. Commun. Mob. Comput.* **2009**, *10*, 843–856. [CrossRef]
12. Chakravarty, S.; Portokalidis, G.; Polychronakis, M.; Keromytis, A.D. Detection and analysis of eavesdropping in anonymous communication networks. *J. Inf. Secur.* **2015**, *14*, 205–220. [CrossRef]
13. Mehta, K.; Liu, D.; Wright, M. Location Privacy in Sensor Networks Against a Global Eavesdropper. In Proceeding of the 20th IEEE International Conference on Network Protocols, Beijing, China, 16–19 October 2007; pp. 314–323.
14. Ghazal, B.; Al-Khatib, K. Smart Home Automation System for Elderly, and Handicapped People using XBee. *Int. J. Smart Home* **2015**, *9*, 203–210. [CrossRef]
15. Song, Y.; Cai, Q.; Nie, F.; Zhang, C. Semi-Supervised Additive Logistic Regression: A Gradient Descent Solution. *Tsinghua Sci. Technol.* **2007**, *12*, 638–646. [CrossRef]
16. Van Kasteren, T.; Noulas, A.; Englebienne, G.; Se, B. Accurate activity recognition in a home setting. In Proceedings of the International Conference on Ubiquitous Computing, Seoul, South Korea, 1–9 September 2008.
17. Park, H.; Park, T.; Son, S.H. A Comparative Study of Privacy Protection Methods for Smart Home Environments. *Int. J. Smart Home* **2013**, *7*, 85–94.
18. Siano, P.; Graditi, G.; Atrigna, M.; Piccolo, A. Designing and testing decision support and energy management systems for smart homes. *J. Ambient Intell. Humaniz. Comput.* **2013**, *4*, 651–661. [CrossRef]
19. Qian, T.F.; Zhou, J.; Wang, M.F.; Cao, C. Analysis of energy consumption in wireless sensor networks. *Comput. Tech. Appl. Prog.* **2007**, *3*, 1710–1714.

![future internet logo] *future internet*

MDPI

Article

An Anonymous Offline RFID Grouping-Proof Protocol

Zhibin Zhou [1,2,†], **Pin Liu** [3,†], **Qin Liu** [4,†] and **Guojun Wang** [5,*,†]

1   School of Information Science and Engineering, Central South University, Changsha 410083, China; zzbzm1031@gmail.com
2   College of Physics and Information Science, Hunan Normal University, Changsha 410012, China
3   School of Information Science and Engineering, Central South University, Changsha 410083, China; jiandanglp@csu.edu.cn
4   College of Computer Science and Electronic Engineering, Hunan University, Changsha 410082, China; gracelq628@hnu.edu.cn
5   School of Computer Science and Educational Software, Guangzhou University, Guangzhou 510006, China
*   Correspondence: csgjwang@gmail.com
†   These authors contributed equally to this work.

Received: 29 November 2017; Accepted: 26 December 2017; Published: 1 January 2018

Abstract: As more and more items are tagged with RFID (Radio Frequency Identification) tags, grouping-proof technology is widely utilized to provide a coexistence evidence for a group of related items. Due to the wireless channel used in RFID systems, a security risk exists in the communication between the reader and tags. How to ensure the tag's information security and to generate reliable grouping-proof becomes a hot research topic. To protect the privacy of tags, the verification of grouping-proof is traditionally executed by the verifier, and the reader is only used to collect the proof data. This approach can cause the reader to submit invalid proof data to the verifier in the event of DoP (Deny of Proof) attack. In this paper, an ECC-based, off-line anonymous grouping-proof protocol (EAGP) is proposed. The protocol authorizes the reader to examine the validity of grouping-proof without knowing the identities of tags. From the security and performance analysis, the EAGP can protect the security and privacy of RFID tags, and defence impersonation and replay attacks. Furthermore, it has the ability to reduce the system overhead caused by the invalid submission of grouping-proofs. As a result, the proposed EAGP equips practical application values.

Keywords: grouping-proof; anonymous; elliptic curve cryptography

1. Introduction

RFID grouping-proof technology is a mechanism that can prove a group of tagged items appeared at the same time and the same place [1]. The grouping-proof protocol can be widely adopted to many applications that need coexistence proof to guarantee the items with RFID tags have been scanned simultaneously, such as supply-chain, health care, and evidence in law [2–4]. For example, in logistics management, we can generate a proof to guarantee the integrity of the container and the goods in it by scanning their tags simultaneously. In the intelligent health care environment, we can validate the correctness of the medicine taking through scanning the patients and their unit-dose medications at the same time and place [5]. In the manufacturing field, a manufacturer of aircraft equipment can certify that a certain part always leaves its factories with a safety cap attached by scanning their RFID tags simultaneously.

According to the connection method between the reader and the verifier, there are two different modes: online and offline [4]. The online mode requires a stable connection between the reader and the verifier, such as [6,7]. In this model, the verifier can send and receive messages from a specific tag (via the reader) during the whole protocol execution. This mode has good real-time performance and high security, but the network condition requirement is relatively high. In some application fields, it is difficult to maintain the network connection between the reader and the background. In addition, the consistent network connection should take the energy efficiency into account [8–10]. On the other hand, in the offline mode, the stable connection between the reader and the background is unnecessary; the reader can collect tag information and generate multiple grouping-proofs without the participation of the verifier. After these processes, the reader can finally send these proof data to the verifier. In this vein, the verifier in offline mode does not need to communicate with any specific tag (via the reader), it only needs the connection before and after the generation of grouping-proof. The connection requirement is more flexible during the protocol, however, there are many security problems need to be solved in this mode, which has become the research focus in many works proposed in the state of the art [3,4,11–18].

Figure 1 shows a common offline mode of RFID grouping-proof system. The tags are divided into M groups: $\{\text{Group}_1, \text{Group}_2, \ldots, \text{Group}_M\}$. Each group represents n_i items with RFID tags. The reader receives group information from the verifier and communicates with tags. If it can simultaneously scan all tags in the ith group, the reader generates a grouping-proof $G_i^{(n_i)}$. After all groups are scanned, the reader sends $\{G_1^{(n_1)}, G_2^{(n_2)}, \ldots, G_M^{(n_M)}\}$ to the verifier. The verifier checks these proofs and stores them as a record. In the grouping-proof protocol, the simultaneous scan means all tags are scanned by a same reader in a short time interval.

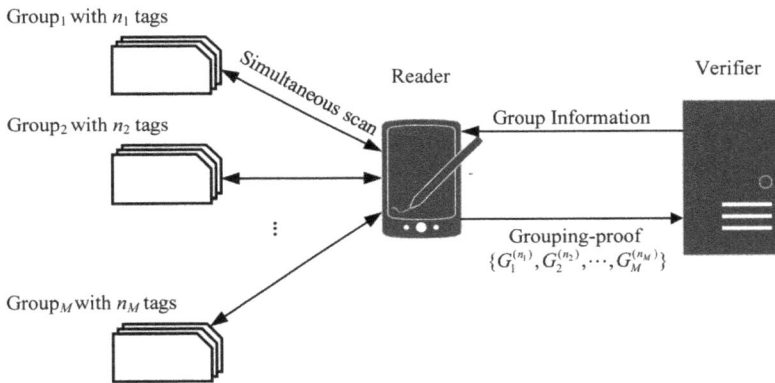

Figure 1. The offline mode of grouping-proof protocol.

1.1. Motivation

In this study, we focus on the offline mode of grouping-proof protocols. There are many works that engage in this mode. At first, the grouping-proof can show the presence of group items as a whole. Note that each single item intends to be sold or transported to other owners. To protect the privacy of these items, the anonymity should be considered as an important security property. In order to do this, the authentication should be anonymous so that any unauthorized third party cannot obtain a tag's identity during the protocol execution. The second point is the secret key distribution. Considering there are a large number of tags in RFID system, the management of secret keys becomes a complicated problem, the use of symmetric encryption schemes is not practical. So the PKI systems are considered. The encryption and decryption in the RSA algorithm need to perform modular exponentiation of great numbers to guarantee security, since the length of the modulus is always larger than 1024 bits, which makes multiplication and division a time-consuming calculation, it is impossible to apply the RSA algorithm in RFID tags in reality. The Ellipse Curve Cryptography (ECC) method is used instead. The point or scalar multiplication is the basic operation for ECC protocols; it is easily performed via repeated group operations which is applicable to low-cost RFID tags. The third problem of offline grouping-proof protocol is that the validity check can only be performed by the verifier. That means the invalid grouping-proof will not be found before submission to the background. This problem greatly reduces the response speed to illegal data. Our solution allows the reader to check the tag's identity before submitting the proof data. However, this solution needs the reader to store the tag's identity, which may bring a potential safety hazard about the tag's privacy information. Therefore, it is essentially necessary to find a way to guarantee the legality of grouping-proof without revealing the secret information of tags.

1.2. Our Contributions

The main contributions of this paper are shown as follows.

(1) We investigate Kang's protocol [19] and provide improvements in key distribution [20], communication overhead, and resistance to impersonation attack and DoP (Denial of Proof) attack.
(2) We establish a scheme to seal the identity of the tag into the grouping-proof message by the group key and session key. So the proof data include two types of tag information: the group member identity and the individual identity.
(3) We propose an ECC based offline anonymous grouping-proof protocol with two tags, denoted as $EAGP^{(2)}$. Based on $EAGP^{(2)}$, we extend the protocol into n tags condition $(n > 2)$, expressed as EAGP. The EAGP has two verification stages. The first stage is used to verify the legality of the tag's group member identity and check the grouping-proof briefly. The second stage is used to verify the identity of the tag and further confirm the grouping-proof.
(4) We carry out the security analysis, performance analysis and correctness proof about the EAGP, and obtain a conclusion that this protocol can resist DoP attack [21] and impersonation attack. It can also protect the tag's information when the reader was compromised. Moreover, EAGP has good scalability in multiple tags condition.

The rest of the paper is organized as follows. An overview of related RFID grouping-proof protocols is presented in Section 2. Section 3 describes the preliminaries of EAGP. Section 4 introduces the Kang's protocol [19]. The system model and definition are described in Section 5. Section 6 shows the EAGP protocol. The security analysis about EAGP are described in Section 7. In Section 8, we provides a performance analysis of our protocols. Section 9 draws a conclusion about this work. The correctness proof about EAGP is described in Appendix.

2. Related Work

The idea of grouping-proof was first introduced in [1], the protocol was called yoking-proof, which only involves two tags coexistence proof in the protocol. Since its introduction, the yoking-proof has evolved to include multiple tags and is now known as the "grouping-proof". In succeeding studies, the grouping-proof protocol is applied in many application fields. In [2–5], the authors used the protocol to generate the medical process evidence for inpatient medication safety. Chien et al. [13] constructed a tree-based tag organization to provide grouping-proof for a complicated system. In addition, there are many other promotions to enhance the security and privacy of this protocol. Burmester et al. in [22] pointed out that there are some problems in grouping-proof protocols: (1) vulnerability to replay attack; (2) unrelated tags can participate in a protocol session, and that the failure can only be found by the verifier; and (3) the protocol does not take the presence of a rogue reader into account. To mitigate these drawbacks, the authors improve the protocol by using group key, proposing the grouping-proof protocol with forward security. Li et al. [16] proposed a yoking-proof protocol with tag anonymous and prove the security within the Universally Composable (U.C.) framework [23]. Cho et al.[18] described a grouping-proof protocol resisted replay attack. In [24], the authors used the code scheme to check the tag information and improve the protocol security.

In [4], the authors analyze the existing grouping-proof protocol, and declared the guidelines for future sound protocols. In order to further improve safety of RFID systems, the application of encryption algorithm is necessary. The work in [25] discussed the feasibility of the ECC in RFID systems. In [26], the authors proposed a RFID chip scheme to support ECC. After that, a RFID mutual authentication protocol based on ECC (ID-Transfer) was proposed [27]. Based on the ID-Transfer, Batina proposed the first grouping-proof protocol based on the ECC in [28] and proved it can provide proof validation and privacy protection in the presence of untrusted tags or reader. The literature [29] showed that Batina's protocol is vulnerable to malicious tracking and proposed the improvement scheme. Kang in [19] further showed that the Batina's protocol is not secure with respect to impersonation attack and they proposed to use the authentication of the reader during the grouping-proof procedure to solve this problem.

3. Preliminaries

In this section, we introduce the ECC and the related hardness problem. The details are described as follows.

3.1. The Ellipse Curve Cryptography

Elliptic curves are algebraic structures that constitute a basic class of cryptographic primitives which rely on a mathematical hard problem. An elliptic curve E over a finite field $\mathbb{F}_q$ with characteristic $q > 3$ can be defined by the Equation (1):

$$y^2 = x^3 + ax + b \tag{1}$$

where $a, b, x, y \in \mathbb{F}_q$ and $4a^3 + 27b^2 \neq 0 \pmod{p}$. The point (x, y) is a point on the elliptic curve. Let P be a fixed point on the curve $E(\mathbb{F}_q)$ with prime order n and k is a large integer scalar in $[1, n-1]$. Due to the hardness of Elliptic Curve Discrete Logarithm Problem [30], it is easy to compute the scalar multiplication $Q = kP$ but hard to find k by knowing only Q and P.

3.2. Elliptic Curve Discrete Logarithm Problem (ECDLP)

ECDLP Definition: Given an elliptic curve E defined over a finite field $\mathbb{F}_q$, a point $P \in E(\mathbb{F}_q)$ of order n, and a point $Q = kP$ where $0 \leq k \leq n-1$, determine k.

The well-known hardness of the ECDLP is crucial for the security of our elliptic curve scheme.

4. Investigation of Kang's Protocol

Literature [19] proposed a grouping-proof protocol based on ECC. The framework of this protocol is shown in Figure 2. Table 1 describes the notations in this protocol.

Table 1. Summary of notations in Kang's protocol.

Notation	Description
P	Base point in the elliptic curve group
k, K	The private/public key of reader
$(s_a, S_a), (s_b, S_b)$	The private/public key of tag A and tag B
y, Y	The private/public key of verifier
$x(T)$	The x-coordinate of point T

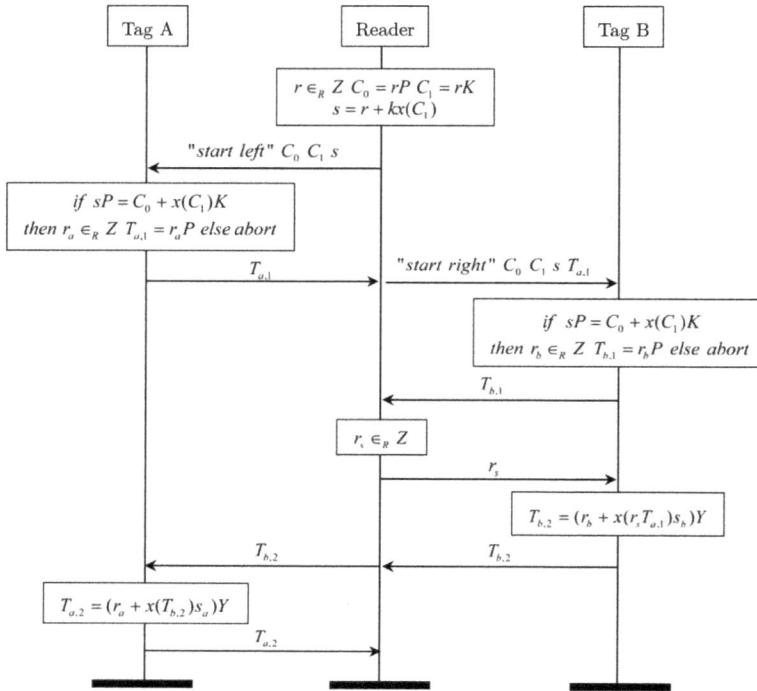

Figure 2. The Kang's protocol.

The protocol has four stages: (1) initialization stage, (2) authentication stage, (3) grouping-proof generation stage, and (4) verification stage. In initialization stage, the server writes the $\{s_a, K, Y\}$ into tag A, the $\{s_b, K, Y\}$ into tag B. The authentication stage is used to authenticate the identity of a reader. It can prevent the reader impersonation attack. In this stage, the reader generates its authentication code $\{C_0 = rP, C_1 = rK, s = r + kx(C_1)\}$ and uses it to identify itself to two tags. Then the reader starts the grouping-proof stage:

(1) According to the random number broadcasted by the reader, tag A generates random number r_a, calculates $T_{a,1}$ and sends it to tag B via the reader.
(2) Tag B calculates $T_{b,1}, T_{b,2}$ and sends $T_{b,2}$ to tag A via the reader.
(3) Tag A calculates $T_{a,2} = (r_a + x(T_{b,2})s_a)Y$ and send it to the reader.
(4) Finally, the reader passes these data as grouping-proof to the verifier for validation.

Kang's protocol uses authentication to solve the impersonation attack, and there are some flaws which need to be pointed out.

(1) The key distribution: in Kang's protocol, tag A and tag B need to store the reader's public key. If the reader is changed, the new public key needs to be written into all the tags. If the amount of tags is very big, the overhead is too serious.
(2) The DoP attack: the reader in Kang's protocol can not validate the proof and is unable to check the legality of tags. If the reader suffered from DoP attack or some unrelated tags taken part into the proof process, before the proof be sent to the verifier, the failure can not be identified immediately which will reduce the system real-time performance.
(3) Communication overhead: the using of authentication stage increases the number of communication times between the tag and the reader, which leads to the additional overhead of communication.

5. The System Model and Security Requirement

5.1. The System Model

In our work, the RFID grouping-proof system is consist of three parts: reader, RFID tags and verifier.

- Tag: the tags in our protocol are passive low-cost devices which have a relative small storage and limited computational capacity. The tags are divided into several groups.
- Reader: the RFID reader is a powerful device which is controlled by an untrusted third party. For security reasons, the privacy information of tag and verifier is unknown to the reader.
- Verifier: an offline trusted third party (TTP) which maintains all the keys and identities of groups.

There are two types of channels in our protocol. The channel between the tag and the reader and the channel between the reader and the verifier. We assume the former is not secure and can be attacked by the adversary. The second channel is secure and the message transferred in this channel cannot be eavesdropped.

5.2. The Adversary Model

In grouping-proof protocols, the adversary has two purposes: (1) forge the grouping-proof which can pass the validation of verifier; and (2) get the privacy information of the reader and tags. According to the attacker described in [23], the adversary in our protocol can completely control the communication channel between the reader and tags, in terms of modifying, delaying and replaying any message in the protocol. In addition, the adversary can also hack the tag and fully control it.

5.3. The Security Requirement of Grouping-Proof System

The security requirements include these parts:

- Anonymity
  The anonymity of tags and readers, which means the adversary cannot get the identity of a tag or a reader by eavesdropping the protocol message.
- Location Privacy
  The adversary cannot track the location of a reader and tags through the protocol messages.
- Resist to replay attack
  The adversary cannot use the message in previous sessions to cheat the reader or tags to generate grouping-proof.
- Defense the DoP Attack
  The adversary cannot use illegal tag involved in the protocol to disturb the proof validation execute by the verifier [21].
- Tag secret information protection
  If the reader is hacked in, the adversary can't use the information stored in it to extract any secret information of tags.

6. Description of EAGP

To overcome the weakness of the grouping-proof protocol which is put forward in [19], we come up with the improvement protocol EAGP.

6.1. EAGP$^{(2)}$

The simultaneous scan is the basic requirement in grouping-proof protocols. To ensure this, the EAGP uses the timeout mechanism to guarantee the tags are scanned by a reader in a very short interval. When the protocol starts, both the reader and tag activate a timer. If a session of grouping-proof do not complete before the timeout, then the protocol is terminated. For simplicity, we assume each group has two tags. Without loss of generality, we assume the verifier can be trusted. The reader and tag are untrusted and can be impersonated or even controlled by an adversary. The notations used in EAGP$^{(2)}$ are summarized in Table 2.

Table 2. Summary of notations in EAGP$^{(2)}$.

Notation	Description
r_s, r_a, r_b	The random number generated by reader, tag A and tag B.
P	The base point on the elliptic curve $E(\mathbb{F}_q)$.
Y, y	The public/private key of Group G.
k_a, k_b	Temporary grouping-proof key of tag A and tag B.
k_{ai}, k_{bi}	Secret key of tag A and tag B.
PK_A, PK_B	Public key of tag A and tag B.
$x(T)$	The x-coordinate of point T.

In EAGP$^{(2)}$, without losing any security characteristics, we cut down the times of communication between the reader and tags to reduce the communication overhead. The proposed protocol consists of three phases: initial phase, grouping-proof generation phase and verification phase.

The descriptions of the protocol are as follows:

6.1.1. Initial Phase

The verifier divides the tag A and tag B into one group, allocates group parameters as: the verifier chooses a random number $y \in \mathbb{Z}$ and computes $Y = y \cdot P$ as its public key. The group's public key Y is stored in the tag, while keeping the private key y. Both tags share their secret keys k_{ai} or k_{bi} with verifier; in addition, the verifier stores the public key PK_A and PK_B. The reader gets the group key y from the verifier.

6.1.2. Grouping-Proof Generation Phase

The framework is demonstrated in Figure 3.

(1) Reader generates a random number r_s, calculates $C_0 = r_s P$, $C_1 = r_s Y$, and $s = r_s + yx(C_1)$. Then, the $\{s, C_0, C_1, r_s\}$ is sent to the tag A along with the message of "*start left*".

(2) Tag A verifies the equation $sP = C_0 + x(C_1)Y$. If it does not hold, the protocol is terminated. Otherwise, it generates a random number k_1, calculates $r_a = x(k_1 P)$, generates the session secret key $k_a = x(Y) \oplus r_a$. Then, it seals its secret key k_{ai} into message m_a as follows:

$$m_a = k_1^{-1}(r_s + k_{ai} \times r_a) \tag{2}$$

Finally, tag A sends $\{m_a, r_a\}$ to the reader.

(3) Reader sends $\{m_a, s, C_0, C_1, r_s\}$ along with the message of "*start right*" to tag B.

(4) Tag B verifies the equation $sP = C_0 + x(C_1)Y$. If it does not hold, the protocol is terminated. Otherwise, it generates a random number k_2, calculates r_b, k_b, m_b, T_b and sends $\{m_b, T_b, r_b\}$ to the reader.

(5) Reader sends the message T_b to tag A.

(6) Tag A calculates $T_a = (m_a + x(T_b)k_a)Y$, and sends it to the reader.

(7) Reader generates the grouping-proof G shown in Equation (3)

$$G = \{m_a, T_a, m_b, T_b, r_a, r_b, s\} \tag{3}$$

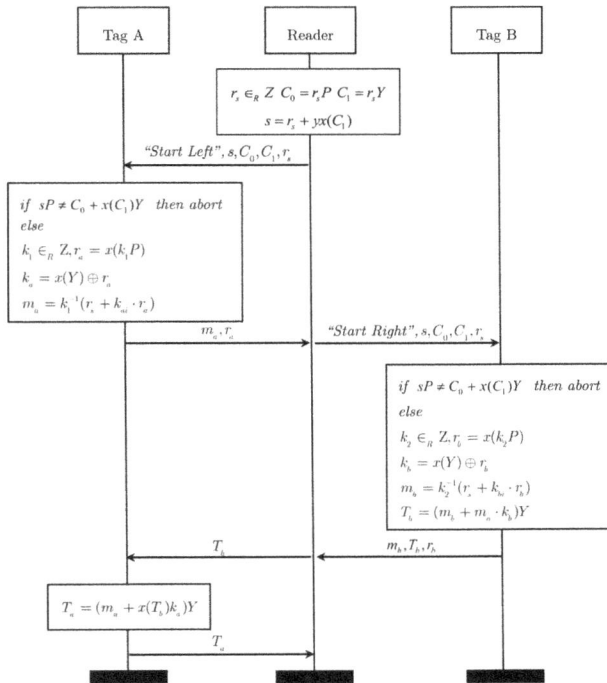

Figure 3. The EAGP.

6.1.3. Verification Phase

There are two steps in the verification phase: (1) Reader verification step, (2) Verifier verification step.

(1) Reader verification step:
Reader calculates $Y' = yP, k_a' = x(Y') \oplus r_a$, $k_b' = x(Y') \oplus r_b$ and validates the Equations (4) and (5):

$$(y^{-1}T_a - m_aP) \times x(T_b)^{-1} = k_a'P \tag{4}$$

$$(y^{-1}T_b - m_bP) \times m_a^{-1} = k_b'P \tag{5}$$

The utilization of group key y can prove that tag A and B belong to the same group and be scanned by the reader simultaneously.

(2) Verifier verification step:
The second verification stage is executed by the verifier to authenticate the tag's identity in grouping-proof. The procedure of tag A is described as follows, the verification of tag B is the same as it:

- Calculate the following equations

$$w = m_a^{-1} \bmod n \tag{6}$$

$$u_1 = s \times w \bmod n \tag{7}$$

$$u_2 = r_a \times w \bmod n \tag{8}$$

$$x_a = x(u_1 P + u_2 PK_A) \tag{9}$$

- If $x_a = r_a$ is valid, the validation is successful, and the verifier stores the proof in the server as a record. Otherwise, the validation fails and the proof is abandoned.

6.2. Extension to n > 2 Tags

In previous description, we assume the group only has two tags, in this section, the EAGP can be extended to multiple tags.

6.2.1. Initial Phase

We describe the group with multiple tags as $G = \{Tag_1, Tag_2, \ldots, Tag_n\}$. The notation of EAGP with n tags can be described by Table 3.

Table 3. Summary of notations in EAGP.

Notation	Description
r_s, r_i	The random number generated by reader and Tag_i.
P	The base point on the elliptic curve $E(\mathbb{F}_q)$.
Y_i, y_i	The public/private Key of Group G.
k_i^t	Temporary grouping-proof key of Tag_i.
k_i, PK_i	Secret/Public key of Tag_i.
$x(T)$	The x-coordinate of point T.

6.2.2. Grouping-Proof Generation Phase

The framework is shown in Figure 4. The solid arrow represents the direct communication, the dotted arrow represents the tag-to-tag communication via the reader.

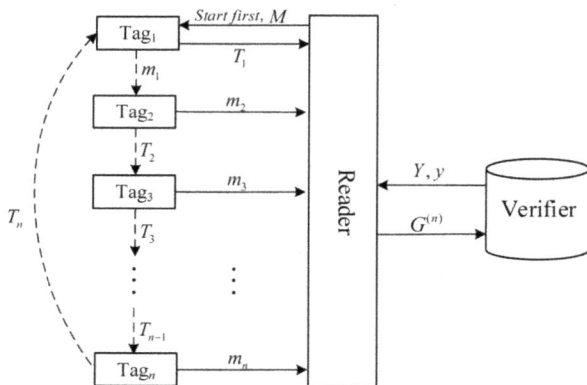

Figure 4. The EAGP with n tags.

(1) Reader selects Tag_1 as the first tag to calculate the grouping-proof. It generates message $M = \{s, C_0, C_1\}$ as Figure 3, and sends it to Tag_1 with the "*Start first*" query.
(2) Tag_1 authorizes the reader, generates message m_1 by Equation (2) and sends it to Tag_2 by the reader.
(3) Tag_2 selects a random number k_2^r, calculates $r_2 = x(k_2^r \cdot P), k_2^t = x(Y) \oplus r_2$, then sends m_2 to the reader, T_2 to Tag_3 via the reader, where

$$m_2 = (k_2^r)^{-1}(r_s + k_2 \times r_2) \tag{10}$$

$$T_2 = (m_2 + m_1 \times k_2^t)Y \tag{11}$$

(4) Tag_3 generates k_3^r and r_3 the same way as Tag_2, calculates m_3, T_3 as below.

$$m_3 = (k_3^r)^{-1}(r_s + k_3 \times r_3) \tag{12}$$

$$T_3 = (m_3 + x(T_2) \times k_3^t)Y \tag{13}$$

Tag_3 sends m_3 to the reader, T_3 to Tag_4 via the reader.

(5) $Tag_i|_{(3<i<n)}$ generates k_i^r and r_i, calculates m_i, T_i as below.

$$m_i = (k_i^r)^{-1}(r_s + k_i \times r_i) \tag{14}$$

$$T_i = (m_i + x(T_{i-1}) \times k_i^t)Y \tag{15}$$

Then Tag_i sends m_i to the reader, T_i to Tag_{i+1} via the reader.

(6)   The last tag Tag_n calculates m_n, T_n, and sends T_n to Tag_1 via the reader.

(7) Tag_1 calculates T_1 by Equation (16), and sends it to the reader.

$$T_1 = (m_1 + x(T_n)k_1^t)Y \tag{16}$$

(8)   The reader generates the grouping-proof $G^{(n)}$ shown in Equation (17).

$$G^{(n)} = \{\, m_1,\ T_1,\ r_1,\ m_2,\ T_2,\ r_2,\ \ldots,\ m_n,\ T_n,\ r_n \,\} \tag{17}$$

6.2.3. Verification Phase

Reader verification step:

The reader verification includes n equations below:

$$k_1^t P = (y^{-1}T_1 - m_1P) \times x(T_n)^{-1} \tag{18}$$

$$k_2^t P = (y^{-1}T_2 - m_2P) \times m_1^{-1} \tag{19}$$

$$k_3^t P = (y^{-1}T_3 - m_3P) \times x(T_2)^{-1} \tag{20}$$

$$\cdots$$

$$k_i^t P = (y^{-1}T_i - m_iP) \times x(T_{i-1})^{-1} \tag{21}$$

$$\cdots$$

$$k_n^t P = (y^{-1} \times T_n - m_nP) \times x(T_{n-1})^{-1} \tag{22}$$

Verifier verification step:

The verifier uses the Equations (6)–(9) to verify the $\{m_1, m_2, m_3, \ldots, m_n\}$ and authenticate the tag's identity.

7. Security Analysis and Comparison

7.1. Security Analysis

7.1.1. The Anonymous of Tag and Reader

During the execution of the protocol, the communication message set can be expressed as $\{r_s, \{m_i, T_i, r_i\}\,|_{i=1,\ldots,n}\}$. Among them, $\{r_i\,|_{i=1,\ldots,n}\}$, r_s are the random numbers generated by tags and reader, while the other messages are calculated from these random numbers. The adversary cannot get any information concerning protocol participants from the communication messages.

7.1.2. The Location Privacy of Tag and Reader

All the messages sent from the EAGP are random numbers or generated from random numbers. In each protocol session, the temporary session key k_i^t and random numbers are different. Adversary cannot figure out the protocol participants by the messages they send. Therefore, it is difficult for the adversary to track any tag or reader, since the locations of readers and tags are protected.

7.1.3. Defense Against DoP Attack

The EAGP adds the reader verification in protocol. When the reader sends the proof to a verifier, the reader can verify the tag's group member identity and proof data before hand. If the adversary does not know the group key, it cannot generate the legal grouping-proof $G^{(n)}$ to satisfy the Equation (21), then it is impossible to cheat the reader to sending invalid grouping-proof to the verifier.

7.1.4. Tag Secret Information Protect

In EAGP, the reader only stores the group's private key y. No tag information is stored in the reader's memory. Even if the adversary gets the group's private key by hacking the reader, it still cannot get any secret information about tag, which makes sure the information security of tags.

7.1.5. Resist to Impersonation Attack

The impersonation attack includes two methods: impersonate tag, and impersonate reader. In the first type, the adversary impersonates the tag, tries to cheat the reader to pass the grouping-proof verification, and further cheats the verifier. In the second type, the adversary impersonates the reader to collect the tag's information, or generates the valid grouping-proof without scanning to the real tag. The attack process is described as follows.

- Impersonate tag
  There are two situations where the adversary impersonate a tag: (1) the adversary does not know any secret key, that means it cannot deduce legal T_i. In this situation, the grouping-proof generated in presence of attack cannot pass the reader validation Equation (21). This attack can be detected before the proof is sent to the verifier, protecting the system from DoP attack. (2) The adversary gets the group's public key Y. From Y, the adversary can deduce the session key k_i^t. Then the adversary can generate the grouping-proof that can satisfy Equations (21). However, due to the lack of tag Tag_i's authentication secret key k_i, to forge the legal m_i need solve the ECDLP described in Section II, thus the probability is negligible. So it is nearly impossible to pass the verifier validation. In conclusion, EAGP can resist the tag impersonation attack in both situations.
- Impersonate reader
  If the adversary impersonate the reader, it needs the group key y to generate s, which is used by tag to authenticate the reader. Without the correct s, the tag will abort the protocol, and the adversary cannot get any information about Tag_i.

From the above, it is difficult for the adversary to impersonate tag or reader. The EAGP can resist impersonation attack.

7.1.6. Resist to Eavesdrop Attack

If the adversary eavesdrop the protocol, the message set it can collect is $\{M, T_i, m_i\}$, all the information is transferred in the ciphertext. Without knowing the secret key of tag, the adversary cannot deduce the tag's identity and forge valid grouping-proof without scanning legal tags.

7.1.7. Resist to Replay Attack

The replay attack denotes when the adversary uses a tag's response to a rogue reader's challenge to impersonate the tag. Suppose the adversary collected the message of Tag_i: $\{m_i^1, r_i^1, r_s^1, s^1, T_i^1\}$ in

EAGP session $p1$, trying to replay these messages in session $p2$ in order to forge a valid grouping-proof including Tag_i while it is absent. The adversary begins the attack as follows:

(1) The adversary sends m_i^1 to the reader.
(2) The adversary sends T_i^1 as T_i^2 to Tag_{i+1} via the reader.
(3) Tag_{i+1} calculates $T_{i+1}^2 = (m_{i+1}^2 + T_i^1 \cdot k_{i+1}^{t2})Y$

Due to the different session $p1$, $p2$, we know $r_s^1 \neq r_s^2$, so $T_i^1 \neq T_i^2$, we get:

$$k_{i+1}^t P \neq (y^{-1} \times T_{i+1}^2 - m_{i+1}^2 P) \times x(T_i^1)^{-1} \tag{23}$$

The grouping proof cannot pass the validation of the reader. EAGP can resist the replay attack.

7.2. Security Comparison

Table 4 lists the comparison of the existing grouping-proof schemes and EAGP. It can be seen from the comparison that the EAGP basically satisfies the security requirements of the grouping-proof protocol.

Table 4. The comparison of grouping-proof protocols.

	Anonymity	Location Privacy	DoP Attack	Tag Information Protect	Tag Impersonation	Reader Impersonation	Replay Attack
Juels [1]	×	×	×	√	×	×	×
Burmester [22]	√	√	×	√	√	×	√
Burmester [24]	√	√	×	×	√	√	√
Batina [28]	√	√	×	√	×	×	×
Chao [29]	√	√	×	√	×	×	√
Lin [31]	√	√	×	√	×	×	×
Kang [19]	√	√	×	√	×	√	√
EAGP	√	√	√	√	√	√	√

8. Performance Analysis

In this section, we analyze the communication overhead of the proposed protocol. The communication overhead denotes the length of the messages transmitted between the reader and tags when they execute the protocol. According to [32], we assume that an elliptic curve with length of 160 bits is used in our schemes. The length of an elliptic curve point is 320 bits. The communicational overhead comparisons about the Kang's protocol [19], EAGP$^{(2)}$ and EAGP are shown in Table 5.

Table 5. The comparison of communication overhead.

		Send		Receive		Total Times
		Total Data (bit)	Transmission Times	Total Data (bit)	Transmission Times	
Kang's	tag A	640	2	1120	2	
	tag B	640	2	1280	2	8
	Reader	2400	4	1280	4	
EAGP$^{(2)}$	tag A	640	2	1280	2	
	tag B	640	1	1120	1	6
	Reader	2240	3	1280	3	
EAGP	Tag_1	640	2	1280	2	
	Tag_i	640	1	1120	1	$2n + 2$
	Tag_n	640	1	1120	1	
	Reader	$1120n + 160$	$n + 1$	$640n$	$n + 1$	

According to the Table 5, we know that the amount of data transferred in EAGP/EAGP$^{(2)}$ and Kang's protocol is very close. However, our protocols reduce the transmission number to six, this will cut down the communication overhead. When the tag number increases to n (EAGP), the transmission data amount of each tag is the same as (EAGP$^{(2)}$), EAGP has good scalability in multiple tags condition.

9. Conclusions

In this paper, we use the ECC as encryption means, cut down the transmission times and propose an offline grouping-proof protocol. In this protocol, the reader can verify the validity of grouping-proof before submitting it to the verifier. The protocol is described in condition of two tags at first (EAGP$^{(2)}$), then we extend it to n tags condition (EAGP). Through the security and performance analysis, EAGP can resist impersonation, DoP and replay attack, protect the security and privacy of tag's secret information.

Acknowledgments: This work is supported in part by the National Natural Science Foundation of China under Grant Numbers 61632009, 61472451 and 61402161, the High Level Talents Program of Higher Education in Guangdong Province under Grant Number 2016ZJ01, the Hunan Provincial Education Department of China under Grant Number 2015C0589, the Hunan Provincial Natural Science Foundation of China under Grant Number 2015JJ3046, the Fundamental Research Funds for the Central Universities of Central South University under Grant Number 2016zzts058.

Author Contributions: Zhibin Zhou contributed to the conception of the study and wrote the paper. Pin Liu contributed significantly to analysis and manuscript preparation; Qin Liu performed the data analyses and wrote the manuscript; Guojun Wang helped perform the analysis with constructive discussions.

Conflicts of Interest: The authors declare no conflict of interest. The funding sponsors had no role in the design of the study; in the collection, analyses, or interpretation of data; in the writing of the manuscript, and in the decision to publish the results.

Appendix A. Correctness Proof of EAGP

Proof of the Correctness about Reader Verify.

For $i = 1$:

If $Y = y \cdot P$ is known, according to Equation (16) we have:

$$y^{-1}T_1 = m_1 P + x(T_n)k_1^t \times P \tag{A1}$$

Then, the right side of Equation (18) can be simplified to $x(T_n)k_1^t P \cdot x(T_n)^{-1} = k_1^t P$. Therefore, the Equation (18) is proved.

For $i = 2$:

According to Equation (11), we have

$$y^{-1}T_2 = (m_2 + m_1 \times k_2^t) \times P \tag{A2}$$

Then, the right side of Equation (19) can be simplified to $m_1 \cdot k_2^t P \cdot m_1^{-1} = k_2^t P$. The Equation (19) is proved.

In a similar way, for $2 < i \leq n$ we have

$$y^{-1}T_i = m_i P + x(T_{i-1}) \times k_i^t P \tag{A3}$$

We put Equation (A3) into Equantion (21), then we can get:

$$(m_i P + x(T_{i-1}) \times k_i^t P - m_i P) \times x(T_{i-1})^{-1} = k_i^t P \tag{A4}$$

The Equantion (21) is proved.

In conclusion, the correctness proof of reader verification is completed. □

Proof of the Correctness about Verifier Authentication.

For the authentication about $Tag_i|_{1 < i \leq n}$, according to Equation (2), we have:

$$k_1 = m_i^{-1}(s + k_i \times r_i) \tag{A5}$$

According to Equations (7) and (8), we have:

$$k_1 = m_i^{-1} \times s + m_i^{-1} \times k_i \times r_i$$
$$= u_1 + u_2 \times k_i \tag{A6}$$

Then, we can obtain:

$$x_i = x(u_1 P + u_2 PK_i)$$
$$= x(u_1 P + u_2 k_i \times P) \tag{A7}$$
$$= x(k_1 P) = r_i$$

The correctness proof of verifier authentication is completed. □

References

1. Juels, A. "Yoking-proofs" for RFID tags. In Proceedings of the IEEE Annual Conference on Pervasive Computing and Communications Workshops, Orlando, FL, USA, 14–17 March 2004; pp. 138–143.
2. Chen, Y.-Y.; Tsai, M.-L. An RFID solution for enhancing inpatient medication safety with real-time verifiable grouping-proof. *Int. J. Med. Inform.* **2014**, *83*, 70–81, doi:10.1016/j.ijmedinf.2013.06.002.
3. Chen, C.-L.; Wu, C.-Y. Using RFID yoking proof protocol to enhance inpatient medication safety. *J. Med. Syst.* **2012**, *36*, 2849–2864, doi:10.1007/s10916-011-9763-5.
4. Peris-Lopez, P.; Orfila, A.; Hernandez-Castro, J.C.; van der Lubbe, J.C.A. Flaws on RFID grouping-proofs. Guidelines for future sound protocols. *J. Netw. Comput. Appl.* **2011**, *34*, 833–845, doi:10.1016/j.jnca.2010.04.008.
5. Zhibin, Z.; Qin, L.; Guojun, W.; Weijia, J. Secure Medication Scheme Using the Grouping-proof Technology. *J. Chin. Comput. Syst.* **2015**, *36*, 2349–2353.
6. Huang, H.; Ku, C. A RFID grouping proof protocol for medicationsafety of inpatient. *J. Med. Syst.* **2009**, *33*, 467–474, doi:10.1007/s10916-008-9207-z.
7. Chien, H.-Y.; Yang, C.-C.; Wu, T.-C.; Lee, C.-F. Two RFID-based solutions to enhance inpatient medication safety. *J. Med. Syst.* **2011**, *35*, 369–375, doi:10.1007/s10916-009-9373-7.
8. Xie, K.; Cao, J.; Wang, X.; Wen, J. Optimal resource allocation for reliable and energy efficient cooperative communications. *IEEE Trans. Wirel. Commun.* **2013**, *12*, 4994–5007, doi:10.1109/TWC.2013.081913.121709.
9. Pizzolante, R.; Carpentieri, B.; Castiglione, A.; Castiglione, A.; Palmieri, F. Text Compression and Encryption through Smart Devices for Mobile Communication. In Proceedings of the 2013 Seventh International Conference on Innovative Mobile and Internet Services in Ubiquitous Computing, Taichung, Taiwan, 3–5 July 2013; pp. 672–677.
10. Castiglione, A.; Palmieri, F.; Fiore, U.; Castiglione, A.; De Santis, A. Modeling energy-efficient secure communications in multi-mode wireless mobile devices. *J. Comput. Syst. Sci.* **2015**, *81*, 1464–1478, doi:10.1016/j.jcss.2014.12.022.
11. Sundaresan, S.; Doss, R.; Zhou, W. Offline grouping proof protocol for RFID systems. In Proceedings of the 2013 IEEE 9th International Conference on Wireless and Mobile Computing, Networking and Communications (WiMob), Lyon, France, 7–9 October 2013; pp. 247–252.
12. Liu, H.; Ning, H.; Zhang, Y.; He, D.; Xiong, Q.; Yang, L. Grouping-Proofs-Based Authentication Protocol for Distributed RFID Systems. *IEEE Trans. Parallel Distrib. Syst.* **2013**, *24*, 1321–1330, doi:10.1109/TPDS.2012.218.
13. Chien, H.-Y.; Liu, S.-B. Tree-based RFID yoking proof. In Proceedings of the International Conference on Networks Security, Wireless Communications and Trusted Computing, Wuhan, China, 25–26 April 2009; pp. 550–553.
14. Lien, Y.; Leng, X.; Mayes, K.; Chiu, J.-H. Reading order independent grouping proof for RFID tags. In Proceedings of the Intelligence and Security Informatics, Taipei, China, 17–20 June 2008; pp. 128–136.
15. Piramuthu, S. On existence proofs for multiple RFID tags. In Proceedings of the 2006 ACS/IEEE Pervasive Services, Lyon, France, 26–29 June 2006; pp. 317–320.
16. Li, N.; Mu, Y.; Susilo, W.; Varadharajan, V. Anonymous yoking-group proofs. In Proceedings of the Proceedings of the 10th ACM Symposium on Information, Computer and Communications Security, Singapore, 14–17 April 2015; pp. 615–620.

17. Ma, C.; Lin, J.; Wang, Y.; Shang, M. Offline RFID grouping proofs with trusted timestamps. In Proceedings of the 2012 IEEE 11th International Conference on Trust, Security and Privacy in Computing and Communications, Liverpool, UK, 25–27 June 2012; pp. 674–681.

18. Cho, J.S.; Yeo, S.S.; Hwang, S.; Rhee, S.Y.; Kim, S.K. Enhanced Yoking Proof Protocols for RFID Tags and Tag Groups. In Proceedings of the Advanced Information Networking and Applications—Workshops, Okinawa, Japan, 25–28 March 2008; pp. 1591–1596.

19. Kang, H.-Y. Analysis and Improvement of ECC-based Grouping-proof Protocol for RFID. *Int. J. Control Autom.* **2016**, *9*, 343–352.

20. Castiglione, A.; de Santis, A.; Masucci, B.; Palmieri, F.; Castiglione A.; Li, J.; Huang, X. Hierarchical and Shared Access Control. *IEEE Trans. Inf. Forensics Secur.* **2016**, *11*, 850–865, doi:10.1109/TIFS.2015.2512533.

21. Lo, N.-W.; Yeh, K.-H. Anonymous coexistence proofs for RFID tags. *J. Inf. Sci. Eng.* **2010**, *26*, 1213–1230, doi:10.6688/JISE.2010.26.4.4.

22. Burmester, M.; De Medeiros, B.; Motta, R. Provably secure grouping-proofs for RFID tags. In *International Conference on Smart Card Research and Advanced Applications*; Springer: Berlin/Heidelberg, Germay, 2008; pp. 176–190, ISBN 978-3-540-85892-8.

23. Canetti, R. Universally composable security: A new paradigm for cryptographic protocols. In Proceedings of the 42nd IEEE Symposium on Foundations of Computer Science, Las Vegas, NV, USA, 14–17 October 2001; pp. 136–145.

24. Burmester, M.; Munilla, J. An Anonymous RFID Grouping-Proof with Missing Tag Identification. In Proceedings of the 10th IEEE International Conference on Radio-Frequency Identification, 3–5 May 2016, Orlando, FL, USA, 2016; pp. 3–5.

25. Wolkerstorfer, J. Is elliptic-curve cryptography suitable to secure RFID tags. In Proceedings of the Handout of the Ecrypt Workshop on RFID and Lightweight Crypto, Graz, Austria, 14–15 July 2005.

26. Batina, L.; Guajardo, J.; Kerins, T.; Mentens, N.; Tuyls, P.; Verbauwhede, I. An Elliptic Curve Processor Suitable For RFID-Tags. *IACR Cryptol. ePrint Arch.* **2006**, *2006*, 227.

27. Lee, Y.K.; Batina, L.;Verbauwhede, I. Untraceable RFID authentication protocols: Revision of EC-RAC. In Proceedings of the RFID, 2009 IEEE International Conference, Orlando, FL, USA, 27–28 April 2009; pp. 178–185.

28. Batina, L.; Lee, Y.K.; Seys, S. Extending ECC-based RFID authentication protocols to privacy-preserving multi-party grouping proofs. *Pers. Ubiquitous Comput.* **2012**, *16*, 323–335, doi:10.1007/s00779-011-0392-2.

29. Lv, C.; Li, H.; Ma, J.; Niu, B.; Jiang, H. Security Analysis of a Privacy-preserving ECC-based Grouping-proof Protocol. *J. Converg. Inf. Technol.* **2011**, *6*, 113–119, doi:10.1.1.464.3789.

30. Menezes, A. *Evaluation of Security Level of Cryptography: The Elliptic Curve Discrete Logarithm Problem (ECDLP)*; University of Waterloo: Waterloo, ON, Canada, 2001.

31. Lin, Q.; Zhang, F. ECC-based grouping-proof RFID for inpatient medication safety. *J. Med. Syst.* **2012**, *36*, 3527–3531, doi:10.1007/s10916-011-9757-3.

32. He, D.; Kumar, N.; Chilamkurti, N.; Lee, J.H. Lightweight ECC Based RFID Authentication Integrated with an ID Verifier Transfer Protocol. *J. Med. Syst.* **2014**, *38*, 116, doi:10.1007/s10916-014-0116-z.

future internet

MDPI

Article

Push Attack: Binding Virtual and Real Identities Using Mobile Push Notifications

Pierpaolo Loreti , Lorenzo Bracciale *and Alberto Caponi

Electronic Engineering Department, University of Rome Tor Vergata, 00173 Rome, Italy;
pierpaolo.loreti@uniroma2.it (P.L.); alberto.caponi@uniroma2.it (A.C.)
* Correspondence: lorenzo.bracciale@uniroma2.it; Tel.: +39-06-7259-7440

Received: 20 December 2017; Accepted: 29 January 2018; Published: 31 January 2018

Abstract: Popular mobile apps use push notifications extensively to offer an "always connected" experience to their users. Social networking apps use them as a real-time channel to notify users about new private messages or new social interactions (e.g., friendship request, tagging, etc.). Despite the cryptography used to protect these communication channels, the strict temporal binding between the actions that trigger the notifications and the reception of the notification messages in the mobile device may represent a privacy issue. In this work, we present the push notification attack designed to bind the physical owners of mobile devices with their virtual identities, even if pseudonyms are used. In an online attack, an active attacker triggers a push notification and captures the notification packets that transit in the network. In an offline attack, a passive attacker correlates the social network activity of a user with the received push notification. The push notification attack bypasses the standard ways of protecting user privacy based on the network layer by operating at the application level. It requires no additional software on the victim's mobile device.

Keywords: online social network; push notification; privacy

---

1. Introduction

Nearly two-thirds of American adults (64%) currently own a smartphone, and more than 91% of smartphone owners aged 18–29 use social networking at least once a week [1]. Facebook has 1.31 billion monthly mobile active users worldwide (44% of them only log in from a mobile device); Twitter has 252 million monthly mobile active users; Instagram 300 million; while the social dating app Tinder has around 50 million active users per month (source: http://www.statista.com/).

Besides the big players, a myriad of different applications usually reside on people's smartphones, allowing their users to play, interact with other users, watch news, etc. To use services (e.g., online social networking, OSN), users must log in to these applications with their accounts. These form their virtual identities.

A virtual identity cannot always be directly traced to the real identity of an OSN user. Pseudonyms represent a simple and easy-to-use way to provide good anonymity. Virtual identities (together with the information associated: political ideas, sexual preferences, etc.) are apparently separated from real identities. In fact, in online social networks, users voluntarily share personal information with the implicit assumption that it is very difficult to link partial information with a person's real identity [2]. This assumption is often false or questionable, as there are many attacks attempting to establish these links: structural re-identification attacks, inference attacks, information aggregation attacks and re-identification attacks [3]. Usually, these techniques use data mining to exploit the correlation among the different information published by one or more people to infer the publisher's real identity.

A binding between virtual and real identities can lead to several issues in countries that limit civil rights, where people get persecuted for their online activity, when virtual identities reveal very

Future Internet **2018**, *10*, 13

personal or sensitive information or when the real person is a public figure. For this reason, the "real names" policy (the policy that obliges users to identify themselves with their real names, banning the use of pseudonyms) adopted by many online platforms over time brought about several complaints from civil liberties groups. Google apologized and reversed its position (https://plus.google.com/u/0/+googleplus/posts/V5XkYQYYJqy) (July 2014), allowing users to pick pseudonyms. Furthermore, Facebook recently promised to pay more heed to real-name complaints (October 2015) (https://www.eff.org/deeplinks/2015/11/facebooks-new-name-policy-changes-are-progress-not-perfection).

In this work, we present a new attack that expressly targets smartphone users to bind their virtual identities with their real ones. Unlike traditional website/desktop applications, smartphone applications have two unique key features:

- smartphones follow users everywhere; so the real identity is, in a certain sense, known a priori.
- smartphone operating systems implement a push notification service that can be exploited to reveal the virtual identity associated with the device owner as described in this work.

These "push notifications" can be emitted asynchronously by app servers and reach mobile clients, even if the related application is closed or suspended and if the terminal is on standby. Push notifications typically (but not always) create new messages in the notification bar, to keep the user updated on what has happened since the last time the app was opened.

It is not uncommon for mobile applications to need asynchronous communication with a remote server.

Push notifications are extensively used by almost all popular mobile apps and especially by social networking applications, for instance to notify the user about a new social interaction such as a new private message or a new friendship request. Furthermore, browsers are starting to support push notifications from websites, and an IETF working group is outlining Internet Drafts [4].

Typically, push notifications can be selectively disabled through app/OS settings. Most are enabled by default and are thus usually active (Figure 1).

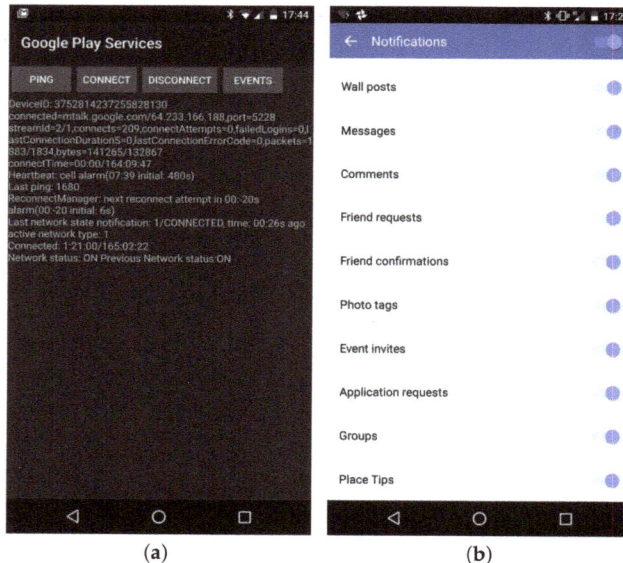

(a)   (b)

Figure 1. Cont.

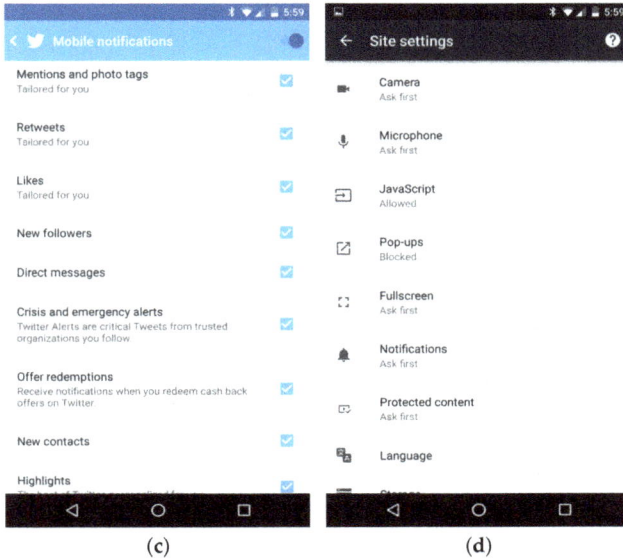

Figure 1. Push notifications: default behavior of popular mobile applications. (**a**) Google Play Service screen showing heartbeat and connection status; (**b**) default Facebook notification settings; (**c**) default Twitter notification settings; (**d**) push notification through Chrome mobile web browser.

If on the one hand, they are usually implemented as asynchronous ciphered messages exchanged through persistent TLS channels, on the other hand, push notifications can be actively triggered. For example, by requesting a social network user's friendship and by eavesdropping the network traffic, they can unveil the presence of a specific user in a specific location, even if the user is using a pseudonym.

1.1. Use Cases

We believe the attack described in this paper can be applied to many scenarios. We propose three reference use cases.

Marketing Campaigns

Customer profiling and tracking comprise an important asset for OSNs where business is mainly targeted to selling specific advertisements. Evaluating the conversion rate from the outcome of a social media campaign (involving virtual identities) into the number of real customers (involving real identities) represents a big issue for business intelligence. A technique that fills this gap will certainly be of interest.

An exemplary use case is determining how many Facebook contacts expressing their will to participate in an event (a concert, show or demonstration) really attend it. Another is identifying the Facebook profile names of the people in a supermarket. This clearly opens several ethical issues. It does though go in the same direction (even if maybe crossing the line) as location analytic services where WiFi-enabled smartphones can be used to indicate customer presence. Indeed, both commercial (e.g., Cisco Meraki CMX [5] or open source (e.g., [6]) software exists that identifies presence, time spent and repeat visits within the range of a WiFi access point by sniffing and analyzing smartphones' 802.11 probe requests.

Working Place

Employee productivity is an important driver for a business' continued viability. However, according to a work survey (http://www.forbes.com/sites/cherylsnappconner/2013/09/07/who-wastes-the-most-time-at-work/#179c48757b3a), 64% of employees visit non-work related websites daily. More than 61 % of workers spend at least one hour per day on these websites. Tumblr, Facebook and Twitter account for the greater share of this category.

Conversely, employee privacy is a fundamental right covered by employment laws in many countries all over the world. Employer surveillance versus employee privacy is indeed an open issue [7].

The privacy attack presented in this paper can reveal the virtual identity of an employee even if they use a pseudonym and a ciphered channel to connect to an OSN. In a workplace, the attack's feasibility increases thanks to the control on the employee's Internet access exercised by the boss, making eavesdropping easy.

Government Monitoring

All major OSNs offer tools for law enforcement and publish governments' information requests for transparency (e.g., Twitter Transparency Reports). At the same time, OSNs have been used to organize riots and protests against governments (2011 Arab Spring Riots) and even to recruit terrorists and to organize attacks (2015 terrorist attack in France and Belgium). This motivates the necessity for governments to reveal users' identities and track the real position of OSN users.

The attack presented in this work can be applied in such scenarios to track the owners of OSN virtual identities. Having control at the ISP level, and without any right to ask for this information from the OSN owner companies, governments can track down dissidents for licit or illicit purposes. The proposed attack can also be used when collaboration with the OSN is not feasible e.g., darknets or where data retention is not available/implemented by the OSN.

1.2. Privacy Implications

The proposed attack is collocated in the wide area of privacy threats on encrypted traffic, often referred to as side-channel information leaks [8]. This area and the related privacy implications have been extensively investigated for the desktop scenario (e.g., [9]). It is recently attracting increasing attention for the mobile scenario. Several works addressed privacy threats and implications because of specific technical and usage characteristics of mobile devices. These include the typical "one app at a time" use [10], the possible privacy leakage from the use of sensors [11], location privacy [12] or, more in general, privacy in mobile environments [13].

This paper, to the best of our knowledge, is the first to deal expressly with using mobile push notifications to bind real and virtual identities.

As already expressed by the authors of [14], the disclosure of virtual and real-world identities could raise severe privacy concerns. Linking them can cause even more damage to an individual's privacy.

Indeed, apparently shielded by pseudonyms and ciphered traffic data (as reported in [2]), people express themselves more freely on politics, religious or personal topics. Linking their virtual to their real identity could then be embarrassing or even harmful (in 2016, Saudi Arabia sentenced a Twitter user to 10 years in prison for publishing 600 tweets "which spread atheism").

Finally, the recent IETF Internet Draft [4] on push notifications also expresses concerns about privacy implications even if from the application's point of view.

1.3. Contributions

This work wants to provide an answer to the following set of questions:

- How do popular applications use push notifications and how can this be exploited to reveal information?

- To what extent is it possible to reveal the presence of a user profile using push notifications?
- What methodologies does the attack use?
- What mitigation techniques are available?

Several works deal with the passive inference of app usage (see Section 6); to the best of our knowledge, this is the first work examining the use of push notifications for an active attack.

This attack is more difficult to conduct than passive measurements, yet it provides a higher grade of precision. We can repeat a test on a given profile many times, lowering the probability of false positives.

This paper is organized as follows: Section 2 presents background information on how notifications and push notifications work; Section 3 presents the details of the proposed attack; Section 4 presents an implementation of the attack; Section 5 presents the results of a measurement campaign to better characterize push notifications in the real world. Finally, Section 6 reviews the present situation, and Section 7 draws conclusions.

2. Background: How Push Notifications Work

2.1. Built-In Push Notification Service

Both Android (https://developers.google.com/cloud-messaging/) and iOS (https://developer.apple.com/notifications/) operating systems provide a built-in service for push notifications. Vendors like Google and Apple offer application developers access to this service through a set of APIs that connect to an infrastructure in charge of delivering a message to mobile devices.

In a typical usage pattern, push notifications start with a third party app server that contacts the vendor server (e.g., gateway.push.apple.com) using a push notification service API. It asks to deliver a given structured message to a certain registered user of an application, identified by a device ID. The push notification server contacts the mobile device passing the push notification to the operating system when the device is online. When offline, the OS stores the notification for a limited period of time, eventually delivering it to the device, when it becomes available. When a push notification reaches the device, it can be typically (but not mandatorily) shown on the notification bar and optionally trigger application code (this behavior strictly depends on the limits of the mobile operating system).

The benefits of implementing push notification at the system level are two-fold: (i) provide an easy way for application developers to have almost synchronous communication between server and application clients; (ii) optimize power consumption by avoiding several persistent connections (one per app), keeping only one shared connection for all apps, multiplexing, en-queuing and possibly stacking all messages directed to the same mobile device.

Although this general schema is valid, the specific implementation of the push notification service varies by the vendor. In what follows, we briefly describe the case of Android and iOS.

Google Cloud Messaging (GCM) (https://developers.google.com/cloud-messaging/) was built in 2012 by Google to provide application short messages (up to 4 kB) from a server to Android devices. Even if the source code of the implementation has not been released (being part of the Google Play Store application), some details were publicly and officially released.

Android push notifications work through one persistent TLS connection opened on port 5228 connecting a device with the GCM servers. This connection is monitored and kept open with periodic heartbeats (e.g., every 240 s) and re-established every time the phone connects to a network. The status of this connection is easily accessible by typing *#*#426#*#* on the phone (Figure 1a).

Apple Push Notification (https://developer.apple.com/library/ios/documentation/NetworkingInternet/Conceptual/RemoteNotificationsPG/Chapters/ApplePushService.html) (APN) works similarly. Each notification is limited to 2 kB. Unlike GCM, APN makes it mandatory for software developers to let their server software notify their user when applications are closed, since there is no way of creating persistent sockets. APN though sets no limit to the number of notifications. APN comes from a ciphered channel (TLS peer-to-peer authentication) on ports 5223 or 443. The source IP addresses are easily identifiable as belonging to Apple's AS (17.0.0.0/8).

2.2. Custom Notification Channels

System notifications also present drawbacks. Besides limits on message size, notification messages can be dropped or delayed. Therefore, some app developers implement their own persistent channel to connect their apps directly with their own servers. See WhatsApp and Facebook on Android.

These channels are usually opened using the system-level push notification service and kept open using keep-alive messages until the user disconnects from the network. In these cases, updates arrive directly from the third party app server to the app on the end user's device, without passing via the vendor infrastructure. These channels transport proprietary messages processed by dedicated code in the apps. They can optionally result (as with the Facebook app) in a new notification shown in the notification bar of the mobile operating system. Conversely, iOS developers must use the Apple built-in notification channel since they cannot build a persistent socket that remains active when the application is closed or suspended.

2.3. Notifications Bar and Erasable Notifications

Applications decide whether to show notifications in the bar. Developers program this choice. For our purposes, the interesting part is that some notifications can hide previous notifications, thus concealing the attack. This behavior is demanded of the application developer. This is a useful feature adopted for instance by Facebook app developers when a friend request is issued and then revoked. In this case, revoking the friendship request causes the previous friendship notification to disappear from the bar. These "erasable notifications" are perfect to perform an attack without alerting the victim. Facebook friend requests are currently erasable on Android, but no longer on iOS.

3. Attack Description

3.1. Adversary Model

The attacker's primary goal is to infer users' personal activities by revealing an exact match between their online (virtual) and real identities. The following reports the attacker's capabilities.

Packet Sniffing

The attacker is modeled as a semi-passive adversary that can silently sniff packets directed to or originated by the victim. To achieve this, the attacker accesses the target user(s)' network A semi-passive rival is analogous to the 'honest-but-curious' or 'semi-honest' adversary model used for security and cryptographic protocols [15]. This attack can be applied to most wireless environments with public access, e.g., WiFi hotspots in neighborhoods, coffee shops, public events, etc. Replication on cellular networks is discussed in Section 3.4. We assume that the interesting traffic is encrypted and that the attacker cannot decrypt transport level packets (e.g., TLS). However, the attack can access lower layer protocol information such as IP source and destination addresses, TCP ports and MAC addresses.

Push Triggering

Besides passive network monitoring, the adversary can trigger push notification messages to the target's personal device using interfaces and actions that are legitimately available on the platform chosen for the attack. For example, a semi-passive adversary can trigger push notification messages by requesting a friendship through an OSN (e.g., Facebook, Twitter, Google+, etc.) or contacting the victim using a messaging platform (e.g., WhatsApp, Viber, WeChat, etc.). The victim must have the related application installed on their mobile device.

Social Data Retrieval

Depending on the type of attack (online, offline), the attacker needs to gather the information on the virtual identities of those users whom they want to bind with their real identities. For example, this information can be the victim's Facebook profile or activity log on the social network.

3.2. On-Line Attack

The proposed attack consists of actively triggering a push notification message to the victim's personal device with the aim of capturing the related data packet sent by the OSN platform to the targeted device. The temporal correlation between the triggering and the sniffed data marks the presence of the victim on the observed network.

The attack scenario in Figure 2 involves the following parties:

- Adversary and victim: The attacker and the targeted virtual identity should be on the same network, for example served by the same WiFi access point or connected to the same Ethernet network.
- Online platform: This usually is the OSN where the victim is registered and represented by the so-called virtual identity. The platform gives the attacker the functionality to trigger push notifications on the user's devices.
- Push notification infrastructure: This is the framework needed to deliver the push notification. It is either platform-specific (e.g., Facebook notification system for Android devices) in the sense that the platform manages push notifications through its own infrastructure or OS-specific when the OS provider manages the notification infrastructure (mandatory for Apple iOS and a common choice for Google Android).

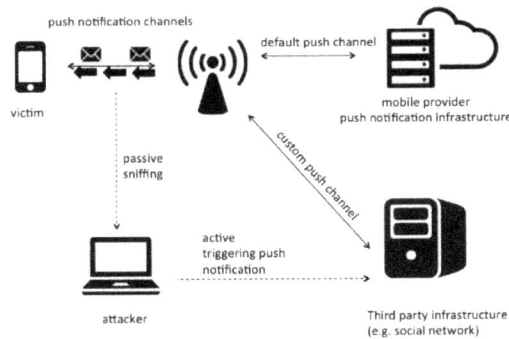

Figure 2. Attack scenario.

The full attack comprises the following steps:

1. Fill a list of possible profile candidates: the attacker targets victims related to a specific location or habits associated with a physical location providing the network access
2. Trigger push notifications to all candidates: the attacker starts to trigger notifications to victims' devices by soliciting online platform events with a pool of bots. This produces push notifications directed to victims' devices that the attacker tries to capture where the targeted user could be connected.
3. Inspect the data over the air, searching for a well-known data pattern: the attacker collects packets related to push notifications from known platforms (e.g., Facebook, Twitter, WhatsApp, etc.) to detect specific size and time patterns that clearly identify the targeted profile.

Victims often cannot detect the attack since attackers can use specific events triggered by the online platform that erase the previous action. An example is requesting the victim's friendship as a first step and later erasing the notification (hopefully before the OS displays it) on the victim's device by removing the friendship request.

Finally, knowing the victim's device's MAC address enables user position estimation by passive triangulation or even by active searching. One can repetitively send ICMP echo requests or ARP requests to the device's MAC address, read the RSSI of the incoming packets and then progressively approach the target.

3.3. Offline Attack

The offline attack has two steps: first, the attacker collects the raw networking data of the monitored location and then correlates this data trace with information crawled from the social network. For example, in Facebook, it is possible to collect timing information on events such as published posts or messages in groups. With this information, the attacker can search the data trace for evidence of the traffic accountable for the push notifications generated by those events.

In this approach, the attacker follows the steps described below:

1   The attacker collects all the packets in a given location related to push notifications from all known OSN platforms (e.g., Facebook, Twitter, WhatsApp, etc.) without actively interacting with victims' profiles.
2   The attacker collects data about the timing of the OSN's events that produced a push notification towards the users.
3   The attacker mixes the data acquired in Steps 1 and 2 to find evidence of the presence of a targeted user in the considered location.

Although the offline attack is conceptually feasible and brings about the discovery of information on the targets, we practically focus the present work on the more viable (and easier to test) online approach. The offline attack is suitable when the attacker has a log trace of the connections performed by several users on different hotspots. An example is having access to logs collected for data retention or security purposes.

3.4. Feasibility of the Attack

Two key aspects impact feasibility: (i) how to fill the list of possible profile candidates; and (ii) how to choose the location for the attack. Both are closely related to the specific context examined for the attack. The former aspect can be (i-a) a "profile check" if the attacker wants to test the presence of a single profile identity (e.g., a given pseudonym) in a given location or (i-b) a "profile search" if the attacker wants to detect if some people belonging to a group are in the surroundings.

In both cases, the WiFi's limited range could hinder the attack. A similar methodology could be used to extend the attack in the case of a wider wireless connection such as LTE (as partially investigated in [16]). In some scenarios, there is no need to eavesdrop the WiFi packets: when the attacker has control of the network backbone serving the WiFi hotspots.

On top of that, offline and online scenarios present specific difficulties. The online attack forces all profile checks in a limited temporal window due to the victim's presence in a given location. Once found, the attacker easily pinpoints the victim using active searching techniques as described above. Conversely, an offline attack requires knowledge of the users' relations (social graph, interaction lists), which sometimes are concealed by the OSN provider according to user privacy settings.

This attack's strong points are that: (i) it assumes no particular setting on the victim's mobile device, (ii) it requires no modification of the standard OSN apps, (iii) it requires no particular equipment (a normal laptop with a WiFi card suffices) and (iv) in some cases (see Section 2), the victim might not even detect it.

As for the differences between the two platforms, we experienced greater technical difficulties in defining the pattern matching on iOS for the reason reported in Section 2. Anyway, Android currently holds 4/5 of the smartphone market.

One can extend the attack to cellular networks using the Paging Message of LTE triggered by incoming network connections. The work in [16,17] adopted this mechanism to track user location. Further investigation on the temporal correlation of LTE paging packets and push notification triggering needs to be carried out to prove the feasibility of the proposed approach on LTE.

3.5. Defense Mechanism

Our technique primarily exploits OSNs' possibility of exploring users' habits by triggering a push notification on the user's device, observing it and exploiting it to infer the victim's physical presence. This exploit allows the attacker to bind a user's virtual identity to a real identity, resulting in a breach of personal privacy. It is thus crucial to provide initial advice to prevent this information linking. Solutions like traffic padding and/or morphing [18,19] can avoid these attacks, but are usually not feasible since it introduces a high amount of network overhead. This can cause several problems especially for not-so-powerful mobile devices: (i) waste of computational resources, (ii) waste of energy and (iii) waste of network traffic that could become a problem for users with limited bandwidth/traffic. Users need more specific solutions to effectively prevent an analysis of push notifications to their mobile devices. Two features allow the enemy to attack:

1. Observing the time between the triggering and the effective reception of the packet on the device. It is highly probable that a push notification packet is received on the user's device shortly after the triggering.
2. Analyzing the size of received packets. The attacker can distinguish between packets related to different push notifications and thus infer the related triggering functionality (e.g., friendship request or comment liked).

A feasible mitigation could be to develop scheduling algorithms focused on avoiding these information leaks. This scheduling should introduce a random delay on message delivery or collect a certain number of push notifications to deliver before triggering the user device with bulk messages. Obviously, this is feasible when the transmission of push notifications is not time-sensitive (i.e., it is not necessary to deliver the friendship request push notification to the user instantly, but it is important in some cases to receive a chat message). The problem of inferring information by observing packet size could be mitigated by hiding information related to the actual size of data in the packet. This could be achieved by producing randomly-sized padded packets for each push notification. This would prevent the attacker from understanding the functionality that triggered the push notification message, i.e., the attacker cannot distinguish if the received message corresponds to a friendship request or a chat message. In addition to the above, in cases in which the push notification packet is delivered by different servers according to the application (e.g., Android GCM), the attacker has one more features to observe and exploit to infer information. Indeed, knowing the packet's source releases information about the specific platform that sent the message and can help the attacker filter out other push notification messages. This could be mitigated by multiplexing push notification traffic like APN already does; only one server delivers all push notifications in the same ciphered data channel.

4. Implementation

We tested the attack using the Facebook Android app, as it is the OSN application with most users. The FB app has the following set of notifications enabled by default: post on wall, private messages, comments, friendship requests, friendship request confirmations, tags, events, application requests, groups. We used the friendship request because it generates a push notification even if the sender and receiver are not friends. In the Android version, it is "erasable" as specified in Section 2; a subsequent friendship request cancellation (revoke) removes any trace in the notification bar, making the attack difficult to spot by the user.

As previously stated, the Android Facebook app does not use the built-in notification system, but sets up, maintains and tears down its own data channel through a persistent Android background service. MQTT [20] messages originated by or directed to Facebook servers transit on these connections. We can easily separate notification data from regular Facebook browsing using reverse DNS lookups, as the former is addressed to servers whose associated name contains the string 'mqtt'. Figure 3 shows the keep alive messages that transit on channel, sent every 60 s to keep the network path up, in an easily recognizable pattern.

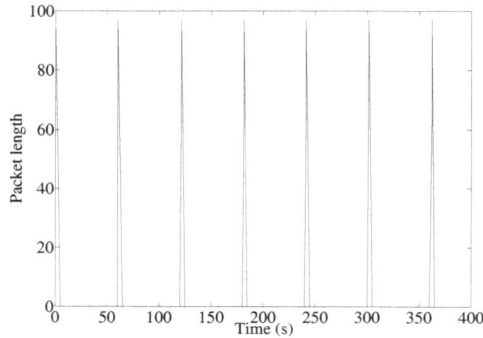

Figure 3. When idle, only heartbeat traffic is visible. This is one packet from client to server and one in the opposite direction, with the respective answers.

When the friendship request arrives, an IP packet of 196 bytes is pushed from the server to the mobile device, followed by a second bigger packet of 615 bytes and a small return packet (85 bytes). Then, a new connection is set up with a different IP address, and a relatively large amount of data is downloaded for a total of 5333 bytes exchanged at the IP level. We suppose these data contain the image and other friendship request meta-data displayed in the notification bar. On the former socket, a small encrypted packet of 195 bytes is sent for the friendship cancellation. This behavior is shown in Figure 4, where we can see a high peak at $t = 153$ corresponding to the friendship request, followed by a smaller peak at $t = 161$, which matches the friendship cancellation request.

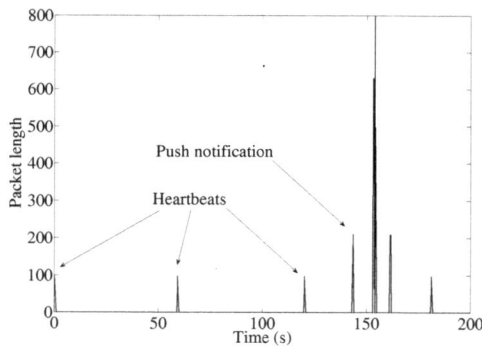

Figure 4. Friendship request and subsequent friendship revoke: received packet lengths over time.

With reference to the steps described in Section 3, we implement the push notification attack in the following way.

Step 1: list possible profile candidates:

To perform the attack, we must leverage prior knowledge of where to search for the list of potential profiles to verify. We consider three different cases: Facebook groups, Facebook participants to an event and Facebook profiles that like a given page. Retrieving the list of users in the first two cases is straightforward as we can use the Facebook graph API to fetch all the members of a group (https://developers.facebook.com/docs/graph-api/reference/v2.4/group/members) or attendees to an event (ttps://developers.facebook.com/docs/graph-api/reference/event/attending/). There are no direct APIs for Facebook pages, probably to protect privacy and prevent spam. Then, we use the Facebook social plugin (https://developers.facebook.com/docs/plugins/page-plugin) designed to embed a widget in a web page showing a random list of people that liked a page. This list gives us a valuable hint about the profiles to check since it includes all users that like a pub, soccer team, political party, campus, conference, course, etc. Using a simple program, we poll this plugin several times to populate the list of possible profile candidates to test.

Step 2: trigger push notifications to all candidates:

Facebook does not provide an API to make friendship requests. By design, Facebook prevents making friendship requests through their Graph API to avoid abuses. Therefore, we use a web browser automation library called Selenium (http://www.seleniumhq.org/). With this library, we build a pool of bots that log in to the social network with different accounts and process the list of candidates by sending and canceling a friendship request to each one, as shown in Figure 5. Facebook limits the number of friendship requests to a few dozen (this quota depends on inner mechanisms: empirically, the number of requests toward a target depends on the relationship between requester and target, e.g., if they share friends or are members of the same group), but increasing the number of fake profiles and inserting a delay, we can easily cope with this limitation. We repeat that we choose to use friendship requests for the reason described above. However, using private messages lets us forge packet length, facilitating pattern matching, and sending messages to a group will trigger a notification to all its members, easing the discovery process.

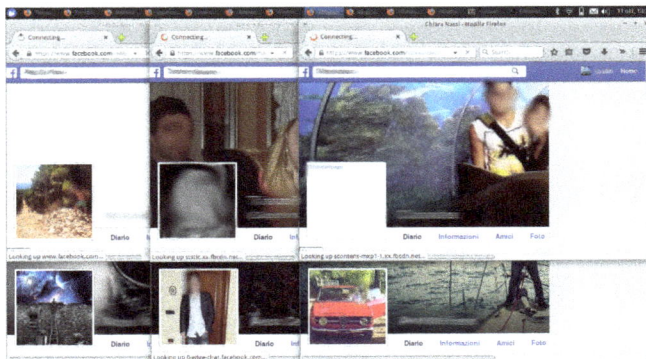

Figure 5. Pool of bots triggering push notifications to a set of profiles.

Step 3: Inspect the data over the air, searching for a well-known data pattern:

We put the WiFi interface in monitor mode and sniff the packets over the air. When we observe an IP packet size of 210 bytes (header + payload) coming from port 443 of a Facebook server whose associated name contains 'mqtt', we mark the user as potentially revealed. A revoke request is 294 bytes long. In this case, the message transits on the Facebook custom notification channel. Empirically, we see that the size of the packet could vary by a few bytes depending on the length of the Facebook ID. If more statistical guarantees of false positives/negatives are needed, one can repeat the experiment or send a pre-determined number of bytes in a private message to a selected subset of users. To cope with WPA encrypted networks with known passwords (as is the case in many public venues), we pre-process the data with a decrypting program (dot11decrypt) that implements monitor mode on 802.11 networks with WPA/WPA2. The decoded traffic is encapsulated using Ethernet frames and written to a tap device.

With iOS devices, the attack can be conducted in a similar way except that notifications no longer come from custom channels, but from the APN. This makes recognition harder as we cannot filter out the packets coming from Facebook IPs. However, we can filter the Apple subnet to separate push notifications from the rest of the traffic. What we have is aggregated traffic of all push notifications addressed to the device as shown in Figure 6. Empirically, friendship requests correspond to packets 590 bytes long.

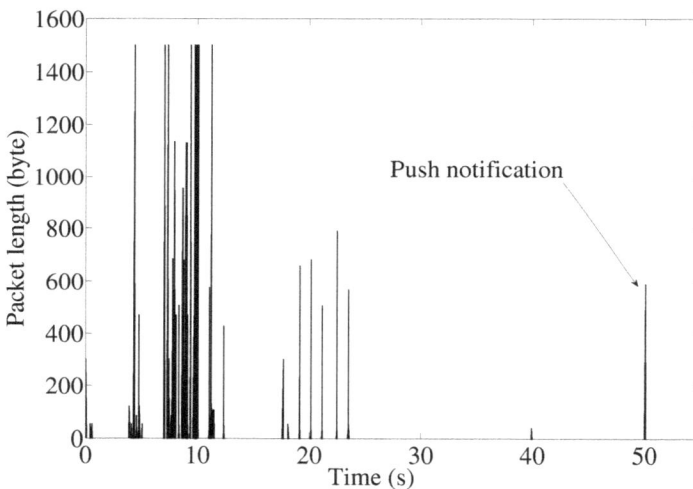

Figure 6. iPhone trace: the push notification is the last spike. The inability to separate Facebook push notifications from other app notifications makes recognition harder than with Android.

The code used in the attack is freely available online (https://github.com/netgroup/push_attack/).

5. Measurements

Being a side channel attack, we are interested in the following information:

- How strict is the temporal binding between the notification triggering and the packet's arrival on the network?
- What is the user identification false positive ratio?

To answer the first question, we repetitively trigger Facebook friendship request notifications and measure the time between the triggering and notification delivery to the mobile device, to estimate the application level round trip time (RTT). Figure 7 reports the empirical distribution of the RTT for a sample of 50 friendship requests. In particular, we recognize the first packet of the push notification that, in turn, triggers other connections responsible for fetching accessory data needed to properly display the notification (e.g., fetching the requester's profile image). Therefore, even if the notification is viewed on the mobile device after several seconds, the packet of interest can be captured after, on average 1.08 s. Despite it being far bigger than the end-to-end network delay (measured at 12 ms), the application RTT is fast enough to enable quite a precise correlation between the triggering of the mobile notification and the packet's detection on the network.

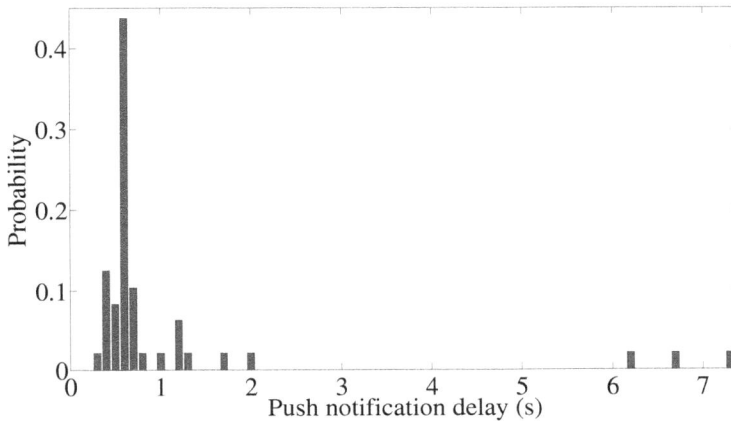

Figure 7. Empirical PDF of round trip time of Facebook friendship request push notification.

As for the second question, we point out that while the attack is performed, victims can also receive push notifications originated by regular application behavior (e.g., a regular friendship request). The attacker can mistake these push notifications for the ones they originated for the attack. In this case, the attacker may erroneously associate a user's identity with the wrong virtual identity, leading to a false positive detection.

Because the attacker knows the length of the push notification that they are searching for, the false positive ratio depends in turn on the distribution of the lengths of push notifications received by users. For instance, a push notification of 196 bytes can be erroneously attributed to a friendship request event.

To estimate the false positive rate and to tune the parameters of the attack, we conduct a measurement campaign in several different environments: a bar, a concert hall, a popular fast-food restaurant, a square, a library. We choose environments where most users connect to the free WiFi with their mobile phones, and we analyze 2 GB of mobile traffic. We extrapolate from the captured data the distribution of the lengths of the push notification messages from Facebook servers to clients. As we can see from Figure 8, the high spike at 66–97 bytes derives from the keep-alive mechanism. The other lengths are quite evenly distributed from 100 to 430 bytes. This fact, on the one hand, suggests the possibility of forging custom length notifications (e.g., private messages) to better cope with a possible false positive detection. On the other hand, the relatively low number of notifications in 2 GB of sniffed traffic demonstrates that application-specific push notifications are quite rare events. This ensures a high degree of precision even in the case of fixed size notifications such as friendship requests.

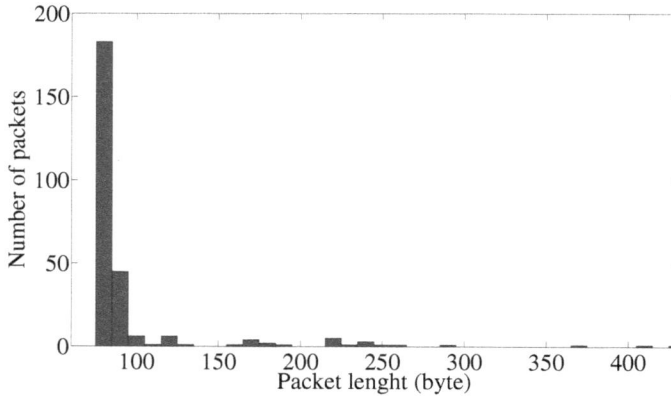

Figure 8. Histogram of the length of Facebook push notifications.

6. Related Work

Mobile OS Vulnerability

Recent works have pointed out how the mobile operative systems present several privacy vulnerabilities that can be exploited by malicious software to retrieve personal data. In some cases, the data are retrieved actively. This is the case of [21], where the authors present a phishing attack based on an installed Trojan application that shows fake notifications to the user.

Sometimes, the apps can just overuse their authorizations to leak personal user data. This problem occurs in open OS, such as Android, where there is no control of the real app behavior. To prevent this abuse, in [22], the authors propose a method for the detection of the personal information leakage. Moreover, since Android Version 6, the user is requested to authorize app permissions at the execution time and not only at the installation time.

As described in Section 3, our approach only needs to trigger push notifications and sniff the network traffic to find a correlation. This is different from other approaches such as [21], as we do not require any control of the victim's smartphone or any special applications installed on their device.

Side Channel Attacks

In addition to basic OS vulnerabilities, the mobile terminals are naturally exposed to attacks based on the wireless traffic analysis that can be performed sniffing the broadcast channel or operating man-in-the-middle attacks. Information encryption is the main technique adopted to protect user privacy in cellular and wireless networks. For example, both Facebook and Twitter communicate with their respective backends using HTTPS and MQTT SSL. However, many works (such as [23–27]) have demonstrated that privacy can be violated by simply analyzing the encrypted traffic, i.e., operating what is often called side-channel information leaks. Our work is in the same field of analyzing ciphered data.

Encrypted traffic patterns can be easily used to detect user activities such as used apps, visited web pages, identification, etc. These attacks operate irrespective of the deployed encryption means and allow one to extract, from the statistical analysis of the generated packet sizes and of their inter-arrival times, valuable confidential information such as the employed applications [28], the application layer protocols [29], the physical devices used [30] or the web page accessed [31]. The side channel attack is more effective in smartphones. For example, in [11], built-in sensors have been utilized to extract application usage patterns. In [9], the authors show that laptop users' activities can be inferred by

MAC addresses using machine learning methods. Moreover, tools for the privacy attacks are freely available: for example, [6] is a distributed tracking and profiling framework that allows one to perform tracking and profiling of mobile users through the use of WiFi. Unlike other side channel attacks on encrypted traffic, the proposed attack is based on active triggering. Consequently, the statistical analysis is devoted solely to finding a well-known pattern in the eavesdropped data. Most other works focus on classifying traffic to infer user data.

Attacks to Anonymity

Both OS vulnerabilities and side channel attacks can be used to acquire personal information of users such their real identities. An example of this attack is reported in [14], where the authors present a technique to link social network profiles with whitepages.com user data. They showed that in some cases, it is possible to correlate the location of the published posts or tweets with the user home location. We differ from [14] as we do not exploit any correlation between public data available on different social network platforms. Furthermore, simple actions such as clicking the Facebook "Like Button" can be used to accurately predict highly sensitive personal data. Some protocols are able to keep un-linkable the identity of their authors with attributes she/he wants to reveal (see [32]). Attacks to privacy in cellular networks have been reported for both GSM [33] and LTE [16]. These attacks are devised to understand the user location exploiting the characteristics of cellular wireless protocols. In particular, [16] shows how it is possible to use a standard LTE terminal to receive the Smart Paging messages generated by the incoming connections (such as VoLTE (Voice over LTE) or application-level messages). They present a methodology similar to the one presented in this paper in one of the proposed attack showing that push notification (together with notifications of Voice over LTE) triggers page notifications that can be sniffed and correlated with the globally unique temporary identifier (GUTI). This analysis complements our work since it is based on active triggering and sniffing on mobile networks to disclose user information. Yet, it has a different goal (location privacy), uses different techniques (smart paging sniffing) and also a different network scenario (cellular network).

7. Conclusions

In this work, we investigated to what extent it is possible to use application-level real-time push notifications to bind users' virtual and real identities and the related privacy concerns. With this aim, we analyzed the behavior and the usage of push notification systems within the most popular mobile social networks and related mobile applications. Most Android applications are easier to attack because developers can use custom/proprietary push notification infrastructures. In these cases, attackers can discover the push notification's source infrastructure. Conversely, iOS applications make use of the built-in OS notification service that multiplexes the notifications of all the applications, making traffic recognition harder. To demonstrate the feasibility of binding by exploiting push notifications, we implemented an example of the attack on the Facebook mobile application. We evaluated the false positive rate through a real-world measurement campaign highlighting the key aspects that characterize the occurrence of push notifications with respect to normal traffic. The push notification attack bypasses the standard technological protections of user privacy, since it works at the application level without the need for additional software on the victim's device. It can be exploited to connect the real to the related virtual identity, causing serious privacy concerns. Future work will focus on assessing the performance of the offline attack using an extended dataset to measure the impact of the attack and on considering different operating systems.

Author Contributions: Lorenzo Bracciale provided the original idea, the description of the attack, and the data analysis; Alberto Caponi performed the security assessment of the attack; Pierpaolo Loreti provides the technological support, the feasibility study and the proof of concept.

Conflicts of Interest: The authors declare no conflict of interest.

References

1. *The Smartphone Difference*; Technical Report; Pew Research Center: Washington, DC, USA, 2015.
2. Guerses, S.; Diaz, C. Two Tales of Privacy in Online Social Networks. *IEEE Secur. Priv.* **2013**, *11*, 29–37.
3. Yang, Y.; Lutes, J.; Li, F.; Luo, B.; Liu, P. Stalking Online: On User Privacy in Social Networks. In Proceedings of the Second ACM Conference on Data and Application Security and Privacy, San Antonio, TX, USA, 7–9 February 2012; pp. 37–48.
4. IETF. Webpush Working Group. Available online: https://tools.ietf.org/wg/webpush/ (accessed on 30 January 2018).
5. *Meraki Whitepaper CMX*; Technical Report; Cisco: San Jose, CA, USA, 2015.
6. Wilkinson, G.; Cuthbert, D. Snoopy: A Distributed Tracking and Profiling Framework. Available online: https://www.sensepost.com/blog/2012/snoopy-a-distributed-tracking-and-profiling-framework/ (accessed on 25 February 2014).
7. Stanton, J.M.; Stam, K.R. The Visible Employee: Using Workplace Monitoring and Surveillance to Protect Information Assets–without Compromising Employee Privacy Or Trust; Information Today, Inc.: Medford, NJ, USA, 2006.
8. Chen, S.; Wang, R.; Wang, X.; Zhang, K. Side-Channel Leaks in Web Applications: A Reality Today, a Challenge Tomorrow. In Proceedings of the 2010 IEEE Symposium on Security and Privacy, Berkeley/Oakland, CA, USA, 16–19 May 2010; pp. 191–206.
9. Zhang, F.; He, W.; Liu, X.; Bridges, P.G. Inferring Users' Online Activities Through Traffic Analysis. In Proceedings of the Fourth ACM Conference on Wireless Network Security, Hamburg, Germany, 14–17 June 2011; pp. 59–70.
10. Wang, Q.; Yahyavi, A.; Kemme, B.; He, W. I know what you did on your smartphone: Inferring app usage over encrypted data traffic. In Proceedings of the 2015 IEEE Conference on Communications and Network Security (CNS), Florence, Italy, 28–30 September 2015; pp. 433–441.
11. Do, T.M.T.; Blom, J.; Gatica-Perez, D. Smartphone Usage in the Wild: A Large-scale Analysis of Applications and Context. In Proceedings of the 13th International Conference on Multimodal Interfaces, Alicante, Spain, 14–18 November 2011; pp. 353–360.
12. Di Luzio, A.; Mei, A.; Stefa, J. Mind Your Probes: De-Anonymization of Large Crowds Through Smartphone WiFi Probe Requests. In Proceedings of the 35th Annual IEEE International Conference on Computer Communications, San Francisco, CA, USA, 10–14 April 2016.
13. Arunkumar, S.; Srivatsa, M.; Rajarajan, M. A Review Paper on Preserving Privacy in Mobile Environments. *J. Netw. Comput. Appl.* **2015**, *53*, 74–90.
14. Alsarkal, Y.; Zhang, N.; Zhou, Y. Linking virtual and real-world identities. In Proceedings of the 2015 IEEE International Conference onIntelligence and Security Informatics (ISI), Baltimore, MD, USA, 27–29 May 2015; pp. 49–54.
15. Goldreich, O. *Foundations of Cryptography: Volume 2, Basic Applications*; Cambridge University Press: Cambridge, UK, 2004.
16. Shaik, A.; Borgaonkar, R.; Asokan, N.; Niemi, V.; Seifert, J. Practical attacks against privacy and availability in 4G/LTE mobile communication systems. arXiv **2015**, arXiv:1510.07563.
17. Stöber, T.; Frank, M.; Schmitt, J.; Martinovic, I. Who Do You Sync You Are?: Smartphone Fingerprinting via Application Behaviour. In Proceedings of the Sixth ACM Conference on Security and Privacy in Wireless and Mobile Networks, Budapest, Hungary, 17–19 April 2013; pp. 7–12.
18. Zhang, F.; He, W.; Chen, Y.; Li, Z.; Wang, X.; Chen, S.; Liu, X. Thwarting Wi-Fi Side-Channel Analysis through Traffic Demultiplexing. *Wirel. Commun. IEEE Trans.* **2014**, *13*, 86–98.
19. Wright, C.V.; Coull, S.E.; Monrose, F. Traffic Morphing: An Efficient Defense Against Statistical Traffic Analysis. In Proceedings of the Network and Distributed System Security Symposium, San Diego, CA, USA, 8–11 February 2009.
20. OASIS. MQTT Specifications. Available online: http://docs.oasis-open.org/mqtt/mqtt/v3.1.1/os/mqtt-v3.1.1-os.html (accessed on 30 January 2018).
21. Xu, Z.; Zhu, S. Abusing Notification Services on Smartphones for Phishing and Spamming. In Proceedings of the 6th USENIX Conference on Offensive Technologies, Bellevue, WA, USA, 6–7 August 2012; p. 1.

Future Internet **2018**, *10*, 13

22. Choi, J.; Sung, W.; Choi, C.; Kim, P. Personal information leakage detection method using the inference-based access control model on the Android platform. *Pervasive Mob. Comput.* **2015**, *24*, 138–149.

23. Herrmann, D.; Wendolsky, R.; Federrath, H. Website Fingerprinting: Attacking Popular Privacy Enhancing Technologies with the Multinomial Naïve-Bayes Classifier. In Proceedings of the 2009 ACM Workshop on Cloud Computing Security, Chicago, IL, USA, 13 November 2009; pp. 31–42.

24. Jiang, T.; Wang, H.J.; Hu, Y.C. Preserving Location Privacy in Wireless Lans. In Proceedings of the 5th International Conference on Mobile Systems, Applications and Services, San Juan, Puerto Rico, 11–13 June 2007; pp. 246–257.

25. Michalevsky, Y.; Nakibly, G.; Schulman, A.; Boneh, D. PowerSpy: Location Tracking using Mobile Device Power Analysis. In Proceedings of the 24th USENIX Security Symposium, Washington, DC, USA, 12–14 August 2015.

26. Sun, Q.; Simon, D.; Wang, Y.M.; Russell, W.; Padmanabhan, V.; Qiu, L. Statistical identification of encrypted Web browsing traffic. In Proceedings of the Security and Privacy, Berkeley, CA, USA, 12–15 May 2002; pp. 19–30.

27. Conti, M.; Mancini, L.V.; Spolaor, R.; Verde, N.V. Can't you hear me knocking: Identification of user actions on Android apps via traffic analysis. *arXiv* **2014**, arXiv:1407.7844.

28. Bernaille, L.; Teixeira, R.; Salamatian, K. Early Application Identification. In Proceedings of the 2006 ACM CoNEXT Conference, Lisboa, Portugal, 4–7 December 2006.

29. Crotti, M.; Gringoli, F.; Pelosato, P.; Salgarelli, L. A statistical approach to IP-level classification of network traffic. In Proceedings of the IEEE International Conference on Communications, Istanbul, Turkey, 11–15 June 2006; Volume 1, pp. 170–176.

30. Kohno, T.; Broido, A.; Claffy, K. Remote physical device fingerprinting. *IEEE Trans. Dependable Secur. Comput.* **2005**, *2*, 211–225.

31. Bissias, G.D.; Liberatore, M.; Jensen, D.; Levine, B.N. Privacy Vulnerabilities in Encrypted HTTP Streams. In Proceedings of the 5th International Conference on Privacy Enhancing Technologies, Cavtat, Croatia, 30 May–1 June 2005; pp. 1–11.

32. Buccafurri, F.; Fotia, L.; Lax, G.; Saraswat, V. Analysis-preserving protection of user privacy against information leakage of social-network Likes. *Inf. Sci.* **2016**, *328*, 340–358.

33. Kune, D.F.; Koelndorfer, J.; Hopper, N.; Kim, Y. Location leaks on the GSM Air Interface. In Proceedings of the NDSS Symposium, San Diego, CA, USA, 5–8 February 2012.

future internet

MDPI

Article

Behavioural Verification: Preventing Report Fraud in Decentralized Advert Distribution Systems

Stylianos S. Mamais *,† and George Theodorakopoulos †

School of Computer Science and Informatics, Cardiff University, 5 The Parade, Roath, Cardiff CF24 3AA, UK; TheodorakopoulosG@cardiff.ac.uk

*   Correspondence: MamaisSS@cardiff.ac.uk; Tel.: +44-29-208-74855

†   Current address: School of Computer Science and Informatics, Cardiff University, 5 The Parade, Roath, Cardiff CF24 3AA, UK.

Received:18 September 2017; Accepted: 16 November 2017; Published: 20 November 2017

Abstract: Service commissions, which are claimed by Ad-Networks and Publishers, are susceptible to forgery as non-human operators are able to artificially create fictitious traffic on digital platforms for the purpose of committing financial fraud. This places a significant strain on Advertisers who have no effective means of differentiating fabricated Ad-Reports from those which correspond to real consumer activity. To address this problem, we contribute an advert reporting system which utilizes opportunistic networking and a blockchain-inspired construction in order to identify authentic Ad-Reports by determining whether they were composed by honest or dishonest users. What constitutes a user's honesty for our system is the manner in which they access adverts on their mobile device. Dishonest users submit multiple reports over a short period of time while honest users behave as consumers who view adverts at a balanced pace while engaging in typical social activities such as purchasing goods online, moving through space and interacting with other users. We argue that it is hard for dishonest users to fake honest behaviour and we exploit the behavioural patterns of users in order to classify Ad-Reports as real or fabricated. By determining the honesty of the user who submitted a particular report, our system offers a more secure reward-claiming model which protects against fraud while still preserving the user's anonymity.

Keywords: advertising; click report; fraud prevention; mobile networks; privacy-preserving advertising

---

1. Introduction

Digital advertising is a form of marketing which capitalizes on the popularity of digital media Publishers such as websites and software applications in order to present consumers with promotional material from retailers who are referred as Advertisers. Ad-Networks such as Google AdSense and Yahoo! AdNet, operate as a middleman between Advertisers and Publishers. Their purpose is to generate revenue by presenting users who browse a Publisher's platform with relevant adverts of the Advertiser's products and services. In more detail, Ad-Networks are granted full control of certain areas within the Publisher's interface for the purpose of featuring adverts. When a user accesses a Publisher's digital domain (whether website or software application), a request is sent to the Ad-Network who selects the most prominent advert from one of the Advertisers and displays it along side the Publisher's content where it may be viewed by the user. The Ad-Network then claims a monetary reward by filing a report to the corresponding Advertiser and afterwards awards a commission to the Publisher for providing his platform.

The ever increasing expansion of the marketing industry makes digital advertising highly profitable for the Ad-Networks and it provides for a tenable source of profit for Publishers that is also beneficial for users who can get services at a lower price. It should therefore come as no surprise that according to eMarketer, the total spending on digital advertising in the US is estimated to exceed

$77 billion for the year 2017 with most of that money being spent specifically on adverts that target mobile users [1].

Despite being highly beneficial for all stakeholders, the Ad-Network advertising model is far from ideal as it is susceptible to fraud. According to a report released by the Investigative Advertising Bureau (IAB) in December 2015, the cost of digital advertising fraud is estimated to be over $8.2 billion with 56% of that (up to $4.6 billion) being attributed to non-human traffic [2]. Ad-Networks attempt to limit the problem through the enforcement of policy-based filtering mechanisms (which detect suspicious traffic), but this approach has not been entirely effective [3]. The Advertisers are the victims of the fraud and although they still benefit from digital ads, they are also forced to tolerate the contemporary state of affairs as they have no other choice but to trust the Ad-Networks.

This research introduces an advert reporting system which utilizes opportunistic networking, blockchain-inspired architecture and client-side processing to entirely substitute Ad-Networks with a community of independent advert-distributors called **Ad-Dealers**. The Ad-Dealers have a physical presence in publicly accessible areas and serve as anonymous communication gateways to Advertisers. Users who appear within proximity of an Ad-Dealer are able to establish a secure connection with the Advertisers in order to transfer information. Opportunistic networking is exploited to further extend the reach of the Ad-Dealers over a social network of mobile users who voluntarily ferry data for each other, while still maintaining privacy through the use of encryption. This enables the system to propagate adverts from the Ad-Dealers to targeted users but also **allows the users to anonymously submit Ad-Reports which they compose themselves**.

Although the majority of users have no immediate benefit from submitting fraudulent Ad-Reports, this does not ensure that a filled report corresponds to real user activity. As users need to remain anonymous, identifying dishonest (malicious) users through traditional methods such as digital signatures is problematic. To address this limitation, our contribution is a mechanism which enables the verification of reports that were submitted by honest users without compromising their identity.

What constitutes a user's honesty for our system is the manner upon which they access adverts on their mobile devices. Dishonest users commit fraud by submitting multiple fake reports over a short period of time while honest users operate under the scope of consumers who view adverts at a balanced pace while engaging in typical social activities. Social activities such as making online purchases, moving through space and interacting with other users are exploited by our system to verify that a report was submitted by a honest user. More specifically, when users perform a particular social task (e.g., visit a designated location or meet other users) they obtain a **Token** which works as proof of the user performing the action at a specific time. To better comprehend this concept, think of a game of scavenger-hunt where players prove to have visited a location by recovering some type of artifact from the area. In a very similar fashion, users of our system collect a series of Tokens and submit them along with their Ad-Reports in the form of a **blockchain-inspired construction** which we term as the Ad-Report Chain (*ARC*). Tokens work like time-stamps which allow Advertisers to calculate the rate at which the submitted reports were composed and identify suspicious activity (dishonest users who create large volumes of reports over short time). This results in the ability to verify that an Ad-Report claim is the outcome of real user activity without need for knowing the identity of the particular user.

Our contributions are as follows:

- To the best of our knowledge, our system is the first to exploit user behavioural patterns (Behavioural Verification) for the purpose of exposing advert fraud.
- Behavioural Verification does not need user identities and therefore preserves privacy.
- A major component of detecting fraudulent behaviour is the rate at which the use creates reports. We simplify the process of calculating this rate compared to the literature. Traditional methods compute this rate by filtering to identify reports that originate from the same source while the approach that we use logs Ad-Reports in a chronological order directly at the user's device and submits them as a single message (*ARC*).

Future Internet **2017**, *9*, 88

The rest of this article is organized as follows. In Section 1.1 we present the scope of our research and in Section 2 we offer insight on some of the related work in the field. In Section 3 we provide preliminary knowledge which is required in order to completely understand the functionality of our system. In Section 4 we provide a high level overview of our model by presenting the system's architecture and formalize the addressed problem. In Section 5 we provide a detailed analysis of our system and offer insight into our approach for solving the addressed problem. Finally, we evaluate our approach in Section 6 and state our conclusion in Section 7.

1.1. Research Scope

The scope of this research is to produce a secure advert reporting mechanism which (1) allows for Ad-Dealers and Publishers to claim service fees from Advertisers, (2) protects all impacted stakeholders (Advertisers, Publishers, Ad-Dealers) from financial fraud by means of preventing the forgery of Ad-Reports and lastly (3) preserves the privacy of the users against all other parties.

2. Related Work

Online fraud is a problem that takes many different forms as expressed in [4]. In our work, we focus on the specific area of advertising fraud which has been continuously expanding over the past years. A number of researchers have contributed different mechanisms with the aim to combat advert fraud. As there are many ways of approaching the issue, their work varies greatly. Some focus on preventing fraudulent reports by detecting and blocking them at their source while others attempt to filter out illegitimate reports by validating their quality after they have been submitted.

The Collaborative Click Fraud Detection and Prevention (CCFDP) [5] model, offers real time click fraud protection capability through the fusion of data (evidence of suspicious behaviour) which is provided by multiple collaborating sources. Three modules are used to independently evaluate reported clicks from both the sever and client side and individually return probabilistic estimates of a click's legitimacy which are combined to produce an overall score. The results of the collaboration are shown to improve the quality assessment of incoming traffic by an average of 10% compared to what is separately achieved by the individual modules, thus allowing the system to identify sources of fraudulent click more accurately and successfully block them.

Rather that filtering our fraudulent clicks, Juels et al. [6] promotes the use of premium clicks which represent reports from users whose legitimacy can be verified through the use of cryptographic credentials, simply known as coupons. Designated websites, referred as attestors, provide their visitors which coupons when they perform specific tasks which are indicative of real user behaviour (e.g., making an online purchase). The coupon can be then attached to future Ad-Reports and works as a form of proof that a particular click was performed by a verified user. The model is implemented in such way that the users' identity is substantially protected against a curious adversary and also offers protection against coupon-replay attacks.

Haddadi [7] argues that click fraud is progressively becoming harder to detect through traditional threshold techniques (identifying multiple reports from the same IP) as BotNet activity is becoming evermore sophisticated through the employment of such means as proxies and distributed attacks. To address the problem, the paper proposes the use of specialized adverts which are called Bluff-ads. Bluff-ads operate as a from of honeypot which allures automated clickers but repels real users. While most adverts are typically targeted at a specific user by being context-specific to a consumer's profile, Bluff-ads are purposely designed to be entirely irrelevant to the user's interests (e.g., an advert for female clothes that is shown to a male user). As Bluff-ads are of no real significance to the targeted user, when they are being clicked may be an indicator of suspicious activity. Although we find this idea to be very creative, we need to remark upon the fact that Bluff-ads are unlike to be adopted as they take valuable space which can be used for real (profitable) ads.

Instead of operating on the server side, FCFraud [8] runs locally on the devices of individual users as a means of preventing them from being part of a BotNet. A BotNet is a group of infected devices which is used to commit click-fraud by generating fake reports without the user's knowledge. The model is incorporated into the operating system as an anti-malware software which monitors submitted click-reports to detect if they correspond to real activity (physical mouse clicks) or have been artificially created by a malicious software. FCFraud is shown to be highly effective at recognizing fake clicks and can easily be implemented in user devices but is only suitable for Advertisers who adopt the PPC (Pay-Per-Click) model.

Faou et al. [9] provide a detailed examination of a click-fraud malware called Boaxxe over a long period. The authors run Boaxxe in a controlled environment and managed to reconstruct a redirection chain which maps the path of different domains that malware follows before been directed to the targeted Advertiser's website. By representing this data in a graph, they were able to identify key actors who have a critical role in the scheme and target them more effectively with the intent of disrupting the malware's operation.

Security within mobile communications has received considerable attention in the context of both cellular [10–13] and opportunistic networks [14–16].

3. Preliminaries

The **Advert Distribution System (ADS)** [17] that we offered in our previous work, combines anonymous download technologies and opportunistic networking in order to deliver adverts to users without exposing their consumer interests. The system that we present in this paper expands the functionality of ADS by enabling users to also submit reports about the adverts that they have viewed.

As both systems share much of the same architectural elements and stakeholders, in this section we provide a fundamental overview of ADS which will be required in order to better comprehend the extended advert reporting system that will be described later on. As depicted in Figure 1, ADS establishes a communication link between **Advertisers**, who are retailers that wish to promote their products and **Users** who represent consumers. The **Ad-Dealers** are local broadcasting stations who serve as communication gateways between users and Advertisers in order to distribute adverts upon request. The role of the Ad-Dealer may be cast to regional entities such as shopping malls, WiFi hotspots and local businesses.

Users run a software client on their smartphone devices that allows them to connect anonymously to Ad-Dealers within their proximity through the use of specialized networking equipment. This allows the mobile clients to freely request and consequently download adverts which are then displayed by **Publishers** who are digital advertising platforms such as websites and mobile apps.

Furthermore, users can also communicate with the Ad-Dealers indirectly through opportunistic networking by forwarding Ad-Requests to an Agent over Bluetooth of WiFi Direct. As shown in the diagram, the Agent physically ferries the Ad-Requests to an Ad-Dealer, collects the requested adverts and conveys them back to the user. In practice, **Agents** are also mobile users who have the additional role of propagating Ad-Requests and adverts on behalf of other users within their social community. This setup benefits members of the community who do not visit Ad-Dealers on a regular basis but can obtain their adverts through other users (Agents) within their social circles.

User privacy against the Ad-Dealers is ensured by the use of anonymous connections (no use of IP address or MAC) and the fact that Agents operate as a form of partially trusted proxy between the users and Ad-Dealers. Ad-Requests are encrypted with a public key which is incorporated into the users' client software and can only be deciphered by the Ad-Dealers who share knowledge of a secret decryption key. The adverts are also encrypted by the Ad-Dealers with a key which is provided to them by the users themselves within the equivalent Ad-Requests. This preserves the privacy of users against Agents who have no access to the content of Ad-Requests nor the adverts and also guarantees security against a rogue Agent who might inject malicious adverts into the network. Security throughout the entire system is also enforced by strong authentication mechanisms as to prevent malicious users from impersonating Ad-Dealers, Agents and other users (In our previous work, we provide a detailed description of how the opportunistic network is established and the mechanisms which are employed to ensure security and anonymity. For this reason, this paper assumes existence of a secure opportunistic link and focuses exclusively on the submission of Ad-Reports).

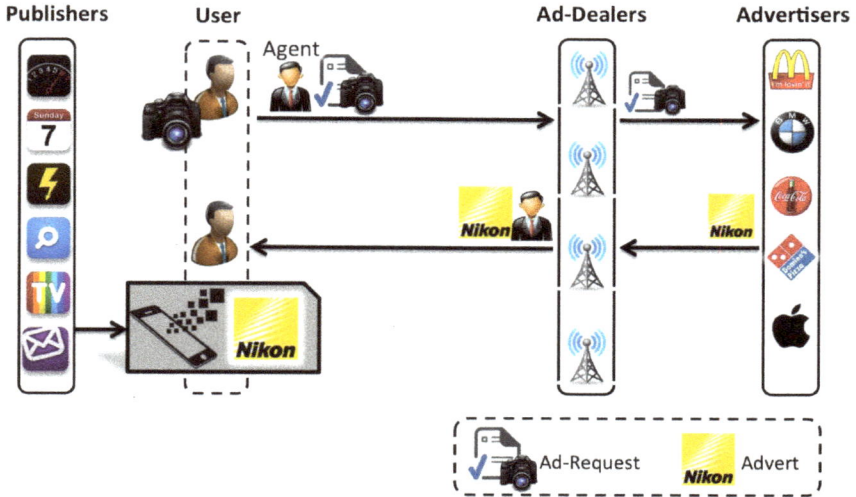

Figure 1. Decentralized advert distribution system over opportunistic network.

4. Model Overview

In this section, we provide a high level overview of our model. We begin by first featuring the stakeholders of our system and detailing their functionality in Section 4.1. We then proceed to define our adversary model in Section 4.2 by contemplating the relationships between the stakeholders in terms of trust and potential threat which enables us to formalize the problem in Section 4.3. Finally, we establish a list of evaluation criteria for our design in Section 4.4.

4.1. System Architecture

As previously mentioned, the advert reporting system that we submit with this work is an extension of our Advert Distribution System (ADS) that we briefly explained in Section 3 and therefore shares much of the same components and architecture. Figure 2 illustrates the architecture of the system and provides a high-level overview of the system's operation which can be divided into three stages with each of them performed under a different sub-protocol later examined in Section 5.3.

The Users, Publishers, Agents, Ad-Dealers and Advertisers represent the same stakeholders as in our earlier ADS system but at the same time have the added functionality of managing Ad-Reports with the help of the Broker. The Broker assumes the role of a trusted representative of the Advertisers to the Ad-Dealers and Publishers. As the Advertisers are too numerous to operate independently while still remaining coordinated, they employ the services of the Broker whose job is to collect and verify the Ad-Reports that are dispatched by the Ad-Dealers.

The user initiates the first stage that is marked in Figure 2 as **(1) Report Form or RF Collection** by composing an RF-Request the purpose of which is to inform the Broker of the user's intention to submit Ad-Reports. The RF-Request is then sent to an Agent who operates as an intermediate node of the opportunistic network. The job of the Agent is to physically transfer the RF-Request to one of the Ad-Dealers, who then forwards to the Broker. Upon receiving the RF-Request, the Broker issues a Request Form (RF) which is sent in the opposite direction: back to the Ad-Dealer, who forwards it to the Agent, so that it may be conveyed back to the requesting user. The RF contains necessary information which the user needs in order to compose and submit his/her Ad-Reports later on.

The second stage, which is marked as **(2) AD Collection**, is essentially the same advert delivery operation which we described in Section 3. Through the use of an Agent, the user sends an Ad-Request message to the Ad-Dealers in order to demand specific adverts. The Ad-Dealer who receives the Ad-Request may then obtain the requested adverts from the appropriate Advertisers and send them back to the user via the same Agent. Note that this operation can take place multiple times via different Agents and Ad-Dealers. Furthermore, it may also take place in combination with the first operation (RF Collection) as one Agent can transfer both the adverts and the RF at the same time.

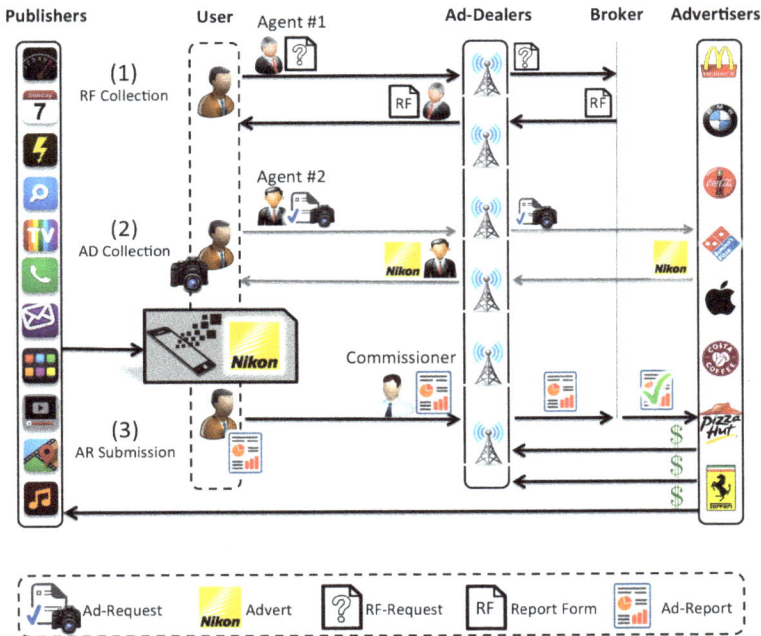

Figure 2. Decentralized advert reporting system over opportunistic network.

The final stage, which is noted as **(3) Ad Report or AR Submission**, takes place after the user has viewed the adverts through a Publisher. To notify the Broker that the user has viewed the adverts, the user composes Ad-Reports and forwards them to a Commissioner with the intent to be delivered to an Ad-Dealer. The receiving Ad-Dealer shares the Ad-Reports with the Broker, who verifies their authenticity and notifies the appropriate Advertisers. Based on the information that is provided within the Ad-Reports, the Advertisers can eventually reward the Publishers who featured the adverts as well as the **two** involved Ad-Dealers (the first that delivered the adverts to the user and the second that submitted the Ad-Reports).

It is evident that the Commissioner serves a very similar function as the Agent. The only difference is that the Agent ferries Ad-Requests and adverts, whereas the Commissioner ferries Ad-Reports. At this point we must note that **the terms Agent and Commissioner are designations that are given to a user based on the type of service that they provide**. Consider an example where a user named Alice sends an Ad-Request to Bob who transfers it to an Ad-Dealer, collects the advert and delivers it back to Alice. Alice then views the advert, generates an Ad-Report that she sends to Charlie who forwards it to the Ad-Dealer. In this particular scenario, the Agent is Bob and Charlie is the

Commissioner. Let us now examine a different scenario where Alice sends her Ad-Request to Danna who collects the adverts for Alice and then Alice composes her Ad-Report that is also sent to Danna who will deliver it to the Ad-Dealer. In this scenario, Danna serves both as Agent and Commissioner. It is clear that there is no practical restriction in regards to who serves as Agent or Commissioner and both responsibilities may be assumed by the same person. As to avoid any confusion, let it therefore be noted that **the term Agent is a designation that is given to the person who delivered an advert to the user and the term Commissioner is a designation that is given to the person that delivered the Ad-Report to the Ad-Dealer and both of these designations may be attributed to the same person**.

4.2. Adversary Model

Our system regards the majority of users as honest since they have no immediate benefit nor the necessary technical knowledge as to undermine the integrity of the system. However, it is possible for an adversary to easily assume the identity of a user without exposing himself since the only precondition for joining the system is to download a mobile client without revealing any identifying information. The main adversary is any group of malicious Ad-Dealers and Publishers who may attempt to commit fraud against the Advertisers by impersonating the identity of a legitimate user in order to submit fictitious Ad-Reports. The Advertisers are therefore willing to accept the authenticity of an Ad-Report only after it has been verified by the Broker whom they consider trustworthy.

The Ad-Dealers and Publishers have no reason to doubt the integrity of the Broker but at the same time are cautious of him as he might alter the content of an Ad-Report as to deprive them of their reword. The Broker therefore serves as a secondary adversary for the Ad-Dealers and Publishers.

One last aspect that needs to be considered is the privacy of the users that needs to be preserved. For this reason, the user considers all other parties, including Agents and Commissioners as honest but curious adversaries and is reluctant to share any personal information that can be associated to his or her identity (Maintaining user privacy when requesting and receiving adverts through an Agent is a matter that we resolved in [17] and for that reason we only focus on the delivery of Ad-Reports. Regardless of this, we do contemplate for all privacy aspects throughout the entirely of the model and thus our evaluation includes a section where we examine how our original advert distribution system is effected).

4.3. Problem Formalization

Given set of Ad-Dealers D, set of Publishers P, set of Users U who submit set of Ad-Reports AR, it should be possible for a Broker b to identify a subset of fictitious reports $\widehat{AR} \subseteq AR$ but at the same time it should not be possible for the Broker b, any Ad-Dealer $d \in D$ or any Publisher $p \in P$ to uncover the identity any user $u \in U$ nor alter the content of any $AR_u \in AR$ without being exposed.

4.4. Evaluation Criteria

Having considered the details of the tackled problem in Section 4.3 and after taking into account the scope of our research in Section 1.1, we dedicate this section to compose an index of system requirements that will serve as the criteria under which the effectiveness and security of our design can be evaluated.

- **Protection against fabricated reports** A malicious user should not be able to commit fraud against the Advertisers by submitting reports which do not correspond to real consumer activity. Fraud prevention is the primary aim of analogous systems and therefore constitutes our main criterion.
- **Report integrity** After an Ad-Report has been sent, it should not be possible for the Ad-Dealers, Commissioners or Broker to alter its content without being exposed. In contrast to other models, our system utilizes a decentralized architecture over insecure channels and should therefore guarantee the integrity of transmitted data.

- **User privacy:** At any given moment, the Broker, Advertisers, Ad-Dealers, Agents and Commissioners should not be individually capable of associating a user's advertising interests to his or her identity. User privacy is an aspect that does not necessarily fall within the scope of fraud prevention systems but is a major concern for our research and is therefore taken into consideration.

- **Reporting effectiveness:** Ad-Reports should include all the necessary information as to ensure that all of the participating stakeholders are able to claim their reward. The Broker should be able to ensure that each report is accounted only once and also there should exist a way for the user to confirm that his or her reports were delivered successfully. The effectiveness of submitted reports is taken as a standard requirement by analogous systems but in the case of our model it needs to be examined in more detail as it may be affected by the additional mechanisms that are used to preserve privacy (opportunistic networks and anonymous submission).

5. Detailed System Analysis

In the following sections we provide a detailed analysis of our system and offer insight into our approach at solving the problem of detecting fake Ad-Reports without the need for knowing the identity of the submitting user. Our approach is based on the notion of behavioural verification, according to which a legitimate user can be verified by his social behaviour. More specifically, when users view adverts on their devices, they generate **Ad-Reports** as featured in Section 5.1.1. At the same time the users also collect a series of **Tokens** when they perform certain social tasks such as visiting specific locations or interacting with other users as explained in Section 5.2. Both the Ad-Reports and the Tokens are composed into an **Ad-Report Chain (ARC)** which is a blockchain-inspired construction that is further analyzed in Section 5.1.2.

As the *ARC* contains both the user's Ad-Reports and Tokens, it can be used by the Broker to verify that the user who created a particular Ad-Report has the same social behaviour as a legitimate user. Furthermore, the Tokens work as time-stamps which allow the Broker to verify that the reports of an *ARC* were created at a paced rate and not in bulk. The *ARC* can then be shared with the Ad-Dealers, Publishers and Advertisers so that the service rewards can be claimed. Information among the system stakeholders is shared through the use of a digital database termed as the **Service Confirmation Board (SC-Board)** illustrated in Section 5.1.3. The *SC-Board* offers additional security mechanisms which allow the user to confirm that his or her reports have been delivered without exposing the user's identity but also enables the Ad-Dealers and Publishers to verify that a submitted *ARC* has not been tampered by the Broker before being published.

5.1. Information Components

In the following sections we introduce the individual informational elements that compose our advert reporting system. More specifically, in Section 5.1.1 we specify the contents of the different types of Ad-Reports and in Section 5.1.2 we outline the way upon which the individual reports of a particular user can be combined into a blockchain-inspired construction that we label the Ad-Report Chain *(ARC)*. Finally, in Section 5.1.3 we present the concept of the Service Confirmation Board *(SC-Board)* that is a medium upon which *ARCs* may be shared with the separate stakeholders of the system.

5.1.1. Ad-Reports

The marketing industry generates profit from online advertising based mainly on three supported revenue models. The first and most popular is called **Cost-Per-Click (CPC)** advertising or Pay-Per-Click (PPC) and is founded on the principle that a reward is attributed when a user clicks on an advert that is displayed on the Publisher's domain. The second most prominent model is referred as **Cost-Per-Impression (CPM)** advertising or Pay-Per-Impression (PPM) and awards money when a user simply views a displayed advert. The third revenue model is known as **Cost-Per-Action (CPA)** or Pay-Per-Action (PPA) where money is awarded when users perform a specific action which is most typically the purchase of a product [18]. The preferences of Advertisers may not be limited to just a single pricing model and for this reason our system can support all of them at the same time by offering three different types of Ad-Reports as listed below.

- **RoV:** Report of View
- **RoC:** Report of Click
- **RoA:** Report of Action

As depicted in Figure 3, all supported Ad-Report types incorporate a sequence number N which indicates the order in which the reports were created. The *Advert Code* is a unique reference number that is sent to the user alongside each advert and can be used for identification. The D_{ID} and P_{ID} respectively accommodate the identities of the Ad-Dealer who distributed the advert and the Publisher who featured it to the user while the *Date* field holds the date and time of the publication.

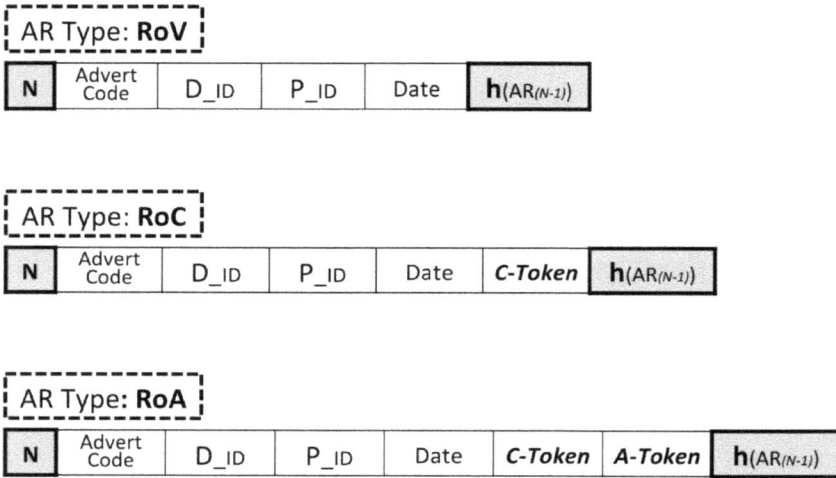

Figure 3. Supported types of Ad-Reports and their contents.

The *C-Token* (or Click-Token), which can be found in the *RoC* and *RoA*, is a sequence of data which can be obtained by the user when the advert is clicked. The *A-Token* (or Action-Token) which is present in the *RoA*, follows a very similar format as the *C-Token* with the main difference being that it is disclosed to the user only after a specific condition has been met (e.g., the user made a purchase or created an account). Each Advertiser periodically generates his own *C-Token* and *A-Token* which are uploaded within his domain. The function which is used for this operation as well as the frequency upon which the two tokens are updated fall udder the responsibility of the respective Advertisers. Ideally, the *C-Token* and *A-Token* should be generated by a cryptographically secure random number generator and as often as practically possible. The rate at which the *C-Token* and *A-Token* are updated influences the system's accuracy of verifying the time that the Ad-Reports were created. More specifically, if the Tokens are updated once every T time units, then the system can verify the time of a user's report with granularity T. It is assumed that the random number generator is secure and that the only feasible way to obtain the *C-Token* and *A-Token* is by downloading them from the locations in which they were uploaded by the particular Advertiser.

More specifically, the *C-Token* is uploaded in the same cuber-space where the user is linked to when clicking on the advert while the *A-Token* is placed in the location to which the user is diverted to when he or she performs a specific action such as making a purchase. Much like the way that web cookies work, the mobile client obtains the *C-Token* and *A-Token* from the Advertiser's website and places them within the Ad-Report as the user is browsing. This enables the Advertisers to verify that a user accessed their website or performed a specific action before creating a *RoC* or *RoA*. Having to obtain the tokens before creating a new Ad-Report, makes the forging of *RoCs* and *RoA* more difficult. To forge a *RoC*, the dishonest user needs to visit an advertiser's web site and to forge a *RoA* requires him to perform an action. More importantly, tokens prevent dishonest users from creating fictitious reports ahead of time as a *RoC* which contains a token C_n could not have been created before the token was made available online. Lastly, $h(AR_{(N-1)})$ contains a hash function digest of each previous Ad-Report that was composed by the same user. This enables the user to link all of the reports that he or she creates in the form of a blockchain-inspired architecture which is analyzed more minutely in Section 5.1.2.

One last thing that needs to be mentioned is the fact that for every Ad-Report, the sequence number N and hash $h(AR_{(N-1)})$ are sent in plaintext form while the remaining fields are encrypted with a public key B_{PuK} which belongs to the Broker and is further clarified in Section 5.3.3.

5.1.2. Ad-Report Chain (ARC) and Integrity Hash (IH)

Rather than dealing with Ad-Reports individually as they are being created, our system enables each user to aggregate multiple Ad-Reports throughout the course of defined period and then submit all of them as a single unit. As it has already been explained in Section 5.1.1, each Ad-Report contains the hash digest of the previous. This enables the user to link several Ad-Reports together in a form that resembles the architecture of a blockchain which is termed the Ad-Report Chain or *ARC* for short.

As shown in Figure 4, the first block of the *ARC* contains an initiating value which is marked as *ARC-ID*. This is hashed to produce $h(ARC - ID)$ that is included in the second block $N = 1$ with each consecutive block following the same arrangement. The $h(ARC - ID)$ essentially works as a unique identifier which also marks the start of a specific *ARC*. The *ARC-ID* is dictated by the Broker and sent to the user within the *RF* (Report Form) as depicted in the same figure. Recall from Section 4.1 that the *RF* (Report Form) is a message that comes as a response to the user's request to file Ad-Reports.

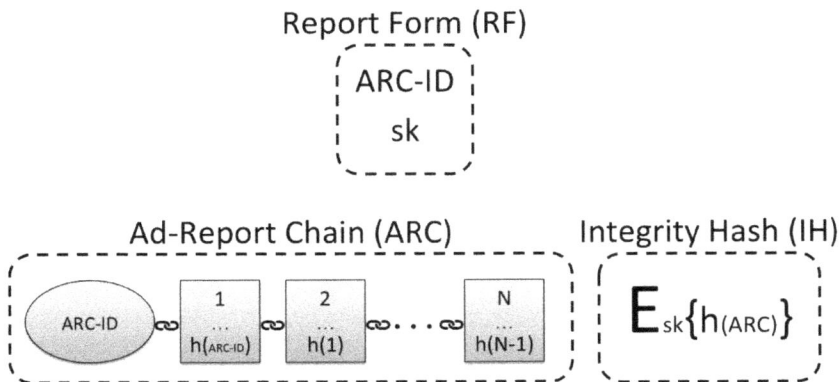

Figure 4. Structural elements of the Ad-Report Chain.

As we can see, the *RF* also contains a cryptographic signing key *sk*. While the *ARC-ID* is used to identify and mark the start of an *ARC*, the *sk* is used to mark the end in such a way that it prevents the removal of addition of blocks. More specifically, when the user completes the creation of the *ARC*, the user produces $h(ARC)$ and then signs it with *sk*. The resulting message, which we can also see in the same figure, is termed as the *IH* (or Integrity Hash) and is sent to the Broker along side the *ARC*.

The Broker can use a secret verification key vk in order to confirm that the IH was created by the user and then determine that the ARC has not been altered by comparing the $h(ARC)$ from within the IH to an $h'(ARC)$ which the Broker computes himself.

5.1.3. Service Confirmation Board (SC-Board)

The Service Confirmation Board (or *SC-Board* for short) is a digital database which serves as an information sharing platform between all of the system's stakeholders. The indexed entries of the *SC-Board* represent Request Forms that have been distributed to users and consist of five fields as shown in Figure 5.

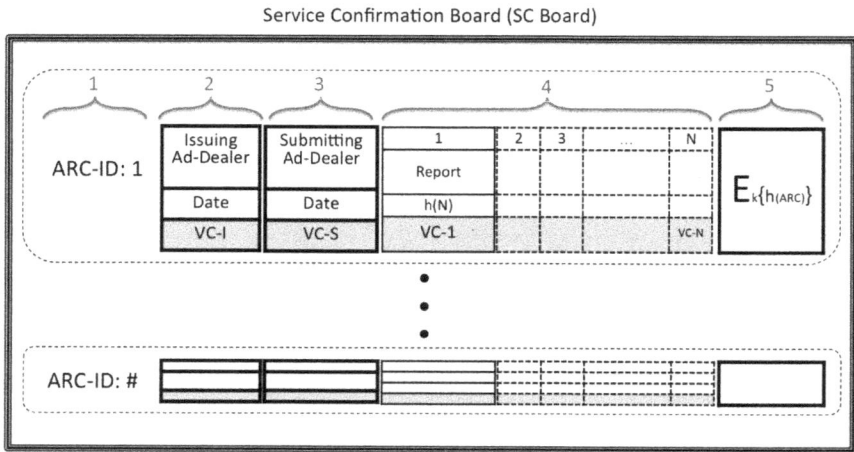

Figure 5. Service Confirmation Board architecture.

The first two fields are filled by the Broker when he issues a new RF and respectively contain the ARC-ID and the name of the issuing Ad-Dealer (the Ad-Dealer who forwarded the user's RF-Request) along with the corresponding date. The remaining fields are completed when the report is submitted with the third field keeping the identity of the submitting Ad-Dealer (the Ad-Dealer who forwarded the user's ARC and IH) and the date of submission while the forth and fifth contain the Ad-Report Chain ARC and Integrity Hash IH.

Indicated in the diagram with a darker shade under the second, third and fourth field, are certain sections which are completed by the issuing Ad-Dealer, the submitting Ad-Dealer, the Broker and the individual Advertisers. These fields serve the purpose of verification checks. In more detail, *VC-I* under the second field is signed by the issuing Ad-Dealer to verify the issue of the new ARC-ID. In a very similar fashion, the submitting Ad-Dealer signs the third field marked as *VC-S* in order to verify the submission of the ARC and confirm the correctness of the hash digests $h(N)$ for each block of the ARC. Recall that in Section 5.1.1, we briefly mentioned that the content of the Ad-Reports are encrypted except for the sequence number N and hash $h(N)$ which are still visible to the submitting Ad-Dealer. While the ARC in the fourth section is published by the Broker after decryption, the submitting Ad-Dealer confirms that the hashes have not been altered by comparing them to his own copy. The individual verification checks, which are marked as *VC-1* to *VC-N* under the ARC, are filled either by the Broker to indicate blocks that have been verified or by the Advertisers to indicate blocks for which the Advertiser has awarded a commission to the Publisher. More details on the exact operation and the reasons behind it are provided in Sections 5.3.3 and 6. Lastly, we need to mention that all fields of the *SC-Board* are visible to Ad-Dealers, Publishers and Advertisers but the first field with the *ARC-ID* also becomes available to the users after submission has been completed. The users only need to have access to the first field in order to verify that their submission has been delivered but cannot see any other information that is published on the SC-Board

5.2. Behavioural Verification

The detection of forged Ad-Reports is a challenging issue because users need to remain anonymous, and anonymity prevents verification through traditional methods such as digital signatures. To resolve this problem, we propose an alternative means of verifying truthful reports while still allowing users to maintain their anonymity. Users can be classified as honest or dishonest based on the manner upon which they create Ad-Reports. As Ad-Reports are rewarded at a low commission (typically at around $1 per 1000 impressions), dishonest users commit fraud on a large scale by generating large volumes of unverifiable Ad-Reports at a rate which is much higher than what is realistically possible for a legitimate consumer. Honest users on the other hand, view adverts at a realistic rate and therefore generate Ad-Reports in a paced manner over a longer period. While they are composing their Ad-Reports, honest users engage in typical social activities such as purchasing goods online, moving through space and interacting with other users. All of these social activities are distinguishing behaviours of honest users which can be exploited to verify their honesty without exposing their identity.

As we already described in Section 5.1.2, the Ad-Reports that are created by the same user are linked together in an *ARC*. The goal is therefore to identify whether the creator of a particular *ARC* is honest or dishonest. We accomplish this by embedding into the *ARC* certain elements (blocks) which reveal the user's social behaviour patterns during the time the Ad-Reports were being created.

5.2.1. Advert Association

Honest users utilize adverts as consumers and are therefore likely to not simply view an advert but to also engage with it by clicking or making a purchase. The act of engaging with an advert can therefore be considered as a typical behaviour of honest users but it also has to be noted that not all honest users engage with adverts in the same rate, and some users do not engage at all. In order to therefore avoid false positives, the system regards the engagement of adverts as an indicator of honesty but the lack of engagement is **not** treated as suspicion of dishonesty. To compensate for users who do not engage adverts, our system exploits other forms of honest behaviour as explained in the following sections.

In Section 5.1.1, we illustrated the available types of Ad-Reports and called attention to the fact that a *RoA* is harder to forge than an *RoC* which is in turn harder to forge than an *RoV* as they contain tokens which are acquired by accessing the Advertiser's website. The *RoCs* and *RoAs* can therefore serve as indicators of honesty as they signify that the user took the time to visit the Advertiser's website. The remaining *RoVs* may not verifiable but they can be validated-by-association since the *ARC* follows the same architecture as a blockchain as shown in Figure 6.

A limitation to this approach lies in the fact that *RoCs* and *RoAs* are designed to be used by Advertisers who support the Cost-Per-Click (CPC) and Cost-Per-Action (CPA) pricing models. This may limit the number of *RoCs* and *RoAs* as it excludes all the Advertisers who only support Cost-Per-Impression (CPM). To overcome this shortcoming, our system utilizes the different types of Ad-Reports (*RoV*, *RoC* and *RoA*) not based on the Advertiser's pricing model but in accordance to the user's engagement with the advert. Regard a simple example where an Advertiser supports the CPM model which means that a simple *RoV* would normally suffice. For the same application however, we can also use a *RoC* or a *RoA* when the user interacts with the advert by clicking or by making a purchase. The commission is still going to be awarded based on the viewing but the use of a more secure Ad-Report will validate the authenticity of the claim as *RoVs* can be forged more easily than *RoCs* and *RoAs*.

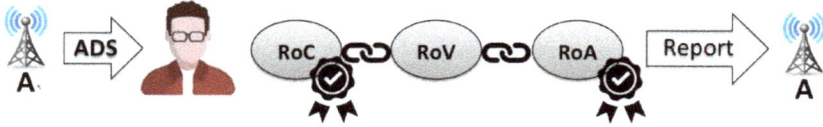

Figure 6. Behavioural verification by advert association.

5.2.2. Time and Location Checkpoint

The use of *RoCs* and *RoAs* enables the identification of honest users in two ways: First, they are indicators of a user who took the time to perform a specific action; second, the *C-Token* and *A-Token* (which are periodically updated) can be used to determine the rate at which the Ad-Reports were created. As not all users engage with adverts regularly enough for this method to be effective on its own, the same principle can be extended by periodically incorporating into the *ARC* some form of Time-Token (*T-Token*) which would signify the time that a particular block was created. One limitation that needs to be considered however, is that this *T-Token* should not be obtained through the Internet, as this would expose the user's IP address and it would also be ineffective since a dishonest user could commit fraud by creating multiple *ARCs* in parallel over a longer period of time.

To overcome this limitation, the *T-Token* is distributed directly from Ad-Dealers in the same way as adverts. The correctness of the *T-Token* is ensured through the synchronization of Ad-Dealers which can easily be achieved with the help of the Broker. This enables the verification of the time that a block of the *ARC* was created but also operates as a location tag. Location tags are data that can be associated with a point in space and time and have appeared in the literature before, in the context of private (cryptographic) proximity testing in [19]. The location tag provides additional proof of the user's honesty as it verifies the user's social behaviour in terms of appearing within the proximity of public locations, where Ad-Dealers are broadcasting, such as shopping malls, cafes and WiFi hotspots.

To further comprehend this notion, consider the following example of attempted forgery. If the *T-Token* were accessible online, a dishonest user $\hat{u}$ could periodically download it and use it to easily verify set $\hat{S} = \{\widehat{ARC_1}, \widehat{ARC_2}, ..., \widehat{ARC_i}\}$ of fictitious *ARCs* over a longer period of time. However, when the *T-Token* is distributed by the Ad-Dealers, it is more difficult for $\hat{u}$ to obtain it since he or she needs to physically travel to the location of the Ad-Dealer and request multiple copies for each of the elements of $\hat{S}$. Furthermore, so as not to raise suspicions, the *T-Tokens* would also need to be requested at a slow rate and preferably from different Ad-Dealers which adds a supplementary layer of difficulty for the adversary.

The *T-Token* is obtained and embedded into a user's *ARC* as follows. When entering the vicinity of an Ad-Dealer, the user sends the hash digest $h(n-1)$ of the last block in the *ARC* and a user-generated encryption key U_{EK}, both encrypted with the Ad-Dealer's public key D_{PuK}. The Ad-Dealer decrypts the message with his private key D_{PrK} and composes a Checkpoint Block (*CB*). As illustrated in Figure 7, the *CB* contains the identity *B* who is the Ad-Dealer that performed the check as well as the received hash $h(n-1)$ and date which are signed with a private key *KB* that belongs to him and is only used for this application. The date serves as the *T-Token* while the signature is proof of the user's location. Before being sent back to the user, the *CB* is first encrypted with the Broker's public key B_{PuK} and then encrypted again with the user's U_{EK} as shown in Equation (1):

$$E_{U_{EK}}[E_{B_{PuK}}[CB]] \tag{1}$$

When the cryptogram is received, the user decrypts it with the corresponding U_{DK} and obtains the $E_{B_{PuK}}[CB]$ which is given a sequence number $N = n$ and is inserted into the user's *ARC* as shown in Figure 7 (All blocks of the *ARC* are encrypted with the Broker's public key B_{PuK} as to ensure privacy against Ad-Dealers and Commissioners. The encryption on the *CB* could had been performed by the user himself but in this case is performed by the Ad-Dealer in order to relieve some of the strain from the user's mobile device. The Ad-Dealer can be trusted with this operation as he has no benefit from providing a defective *CB*).

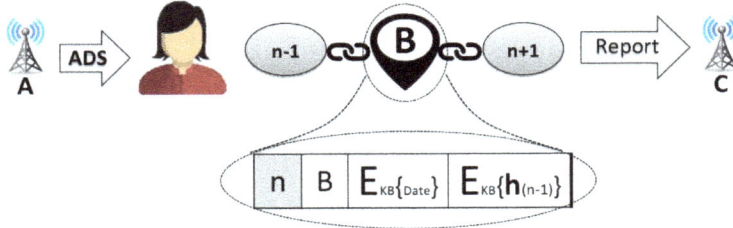

Figure 7. Behavioural verification by checkpoint.

5.2.3. Social Affiliation

The social affiliations between honest users is an additional behaviour which can be exploited to verify the rate that an *ARC* was created. When two users meet, they may exchange the sequence numbers N and hashes $h(n)$ of their last blocks as well as their *ARC-IDs*. The two users can then verify the date and time of the meeting by adding an Affiliation Block *AB* in their respective *ARCs* with each others' information. In order to be valid, the *ABs* which are added to the *ARCs* of both users need to have matching dates and times but this does not require a perfect synchronization between the two users. Mobile applications typically have a recommended refresh rate for adverts that is between 30 s to 120 s while malicious applications generate Ad-Reports at a higher rate. For the purpose of detecting fraud, the time difference between the two users can therefore be tolerant to a margin of a couple of minutes without seriously affecting the system. In the event that two *ABs* do not match because one of the users provided an unrealistically inaccurate date and time (either maliciously or accidentally), the Broker can simply ignore it while relying on other Tokens to validate the particular *ARC*.

Figure 8 illustrates an example where two users A and B have added each others' Affiliation Blocks within their respective *ARCs*. The *AB* which was added by user B, is shown in the diagram to contain a new sequence number n, the hash of the previous block $h(n-1)$, the **Date** of the meeting (as registered by B) and the information that was sent by A which includes her *ARC-ID=ARC1* as well as the sequence number m and digest $h(m)$ of her last block. The date which is added by B works as a *T-Token* which verifies the last block of *ARC1* at a particular time. Notice that *ARC1* and sequence number m are sent encrypted with the Broker's public key B_{PuK} while the $h(m)$ is signed with A's signing key *sk1* which is also used for the creation of the Integrity Hash *IH* as explained in Section 5.1.2. This ensures that B does not lean any information about *ARC1* and is not able to alter $h(m)$.

Through the exchange of *ABs*, the Broker can infer that two *ARCs* were submitted by affiliated users but this does not compromise user privacy in any way. The *ARCs* are submitted anonymously and the Broker has no means of obtaining any information about a particular user's social network nor is he able to identify *ARCs* that were submitted by the same user. One limitation of this verification method lies on the fact that a dishonest user may exchange *ABs* between multiple fictitious *ARCs*. Although plausible, this is prevented by combining all three verification methods as discussed in the following Section 5.2.4.

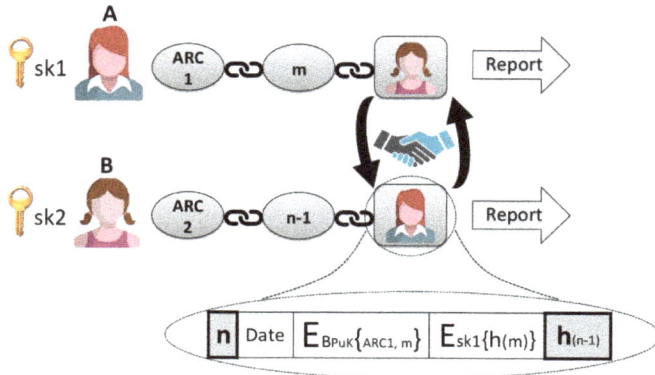

Figure 8. Behavioural verification by social affiliation.

5.2.4. Combined Verification

The individual methods of behavioural verification have certain limitations as the social affiliation approach in Section 5.2.3 is susceptible to fraud by means of creating multiple fictitious *ARCs* while the methods which are described in Sections 5.2.1 and 5.2.2 may not always be practically feasible as they require the user to regularly click on adverts or travel to certain locations.

To compensate for each others' limitations, all three approaches were designed to work in combination. In the example which is provided in Figure 9, user *A* submits an *ARC* which contains multiple Ad-Reports that need to be verified (marked in the figure with exclamation marks). The honesty of *A* is supported by the fact that his *ARC* also contains a verifiable report (either a *RoC* or a *RoA*), a Checkpoint Block from an Ad-Dealer and two Affiliation Blocks.

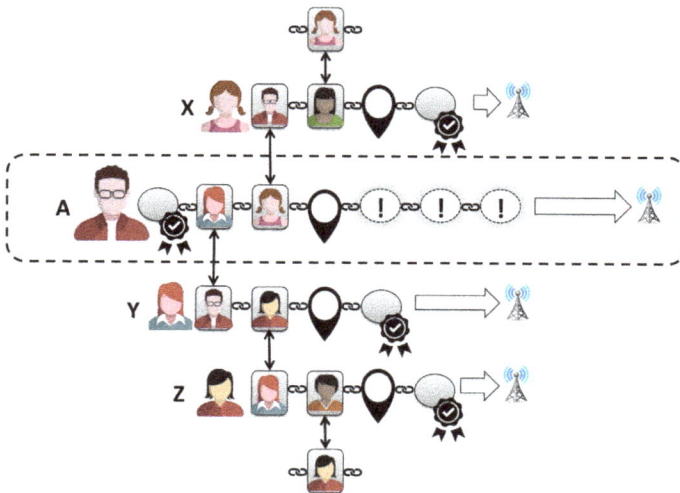

Figure 9. Extended diagram of behavioural verification.

Furthermore, we see that the respective *ARCs* of the two users *X* and *Y* who provided *ABs* for *A* also have verifiable reports, *CBs* and *AB* from other users such as *Z*. As all submitted *ARCs* show indications of social activity, it serves as significant evidence to support the notion that they were composed by different honest users rather than a single dishonest one. For reasons of simplicity, the example just described features only a few verification credentials. However, in a more realistic scenario, the users would likely have multiple credentials which would solidify their verification.

5.3. Protocol Description

The system comprises of three sub-protocols which correspond to the three stages of the system's operation (1) Report Form Collection, (2) AD Collection and (3) Ad Report Submission that were mentioned in Section 4.1. The sub-protocols run sequentially and are detailed in the following sections.

5.3.1. Report Form Collection Sub-Protocol

The Report Form Collection sub-protocol that is depicted in Figure 10 is run when the user needs to acquire a new Report Form.

1. The user calculates a pair of asymmetric keys U_{EK} and U_{DK}. The encryption key U_{EK} is composed into an *RF-Request* and the decrypton key U_{DK} is temporarily stored.
2. The *RF-Request* is encrypted with the Broker's public key B_{PuK} and sent to one of the Ad-Dealers, either directly or via an Agent.
3. Upon receiving the *RF-Request*, the Ad-Dealer forwards it to the Broker.
4. The Broker decrypts the *RF-Request* with his private key B_{PrK} and obtains the user's encryption key U_{EK}. The Broker then creates an entry in the *SC-Board* with a new *ARC-ID* and the identity of the issuing Ad-Dealer and afterwards computes a pair of keys *sk* and *vk*. The Verification Key *vk* is stored securely while the Signing Key *sk* and the *ARC-ID* are composed into a Request Form *RF*.
5. The *RF* is first encrypted with the U_{EK} which was provided by the user and then is returned to the issuing Ad-Dealer who filed the request.
6. The **issuing** Ad-Dealer verifies his identity on the *SC-Board* by providing this signature.
7. The Ad-Dealer forwards the *RF* to the user either directly or via the same Agent who delivered the original *RF-Request*.
8. The user receives the encrypted *RF* and decrypts it with his U_{DK} in order to obtain the *ARC-ID* and *vk*.

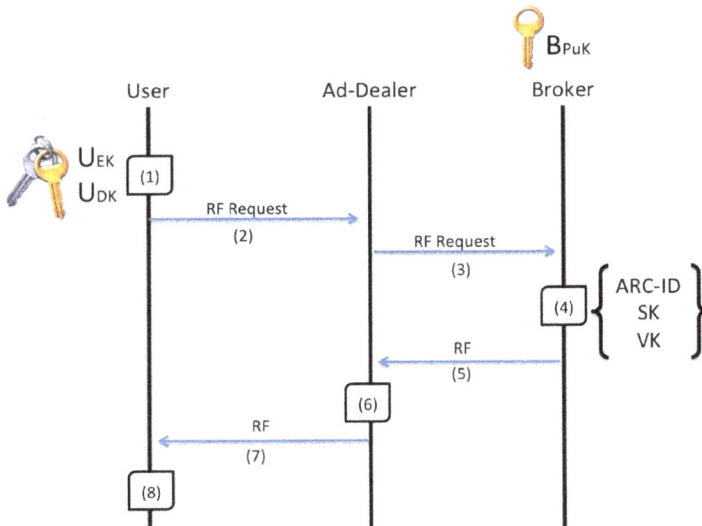

Figure 10. Report Form Collection sub-protocol.

5.3.2. Advert Collection Sub-Protocol

The Advert Collection sub-protocol that is depicted in Figure 11 is used for the acquisition of adverts and may run multiple times as needed.

1. The user computes a pair of asymmetric keys U'_{EK} and U'_{DK}. The user stores the decryption key U'_{DK} and then composes an Advert Request Message (ARM) which contains the encryption key U'_{EK} and his advertising interest (referenced by known interest identifies that are common throughout the system).
2. The ARM is encrypted with a public key D_{PuK} and sent to one of the Ad-Dealers, either directly or via an Agent.
3. The responding Ad-Dealer decrypts the ARM with the corresponding key D_{PrK} which is kept private among all the Ad-Dealers who use the system and obtains U'_{EK} as well as the user's advertising interest.
4. The Ad-Dealer determines which specific adverts would best suit the needs of the user and files a request to the appropriate Advertisers.
5. Upon receiving the Ad-Dealer's request, the Advertisers respond with the adverts in plain form.
6. The Ad-Dealer aggregates all of the received adverts into an Advert Delivery Message ADM.
7. The ADM is first encrypted with the U'_{EK} and then returned to the user either directly or via the same Agent who delivered the original ARM.
8. The user receives the decrypts the ADM with his U'_{DK} in order to obtain the adverts which are stored for future use by the Publishers.

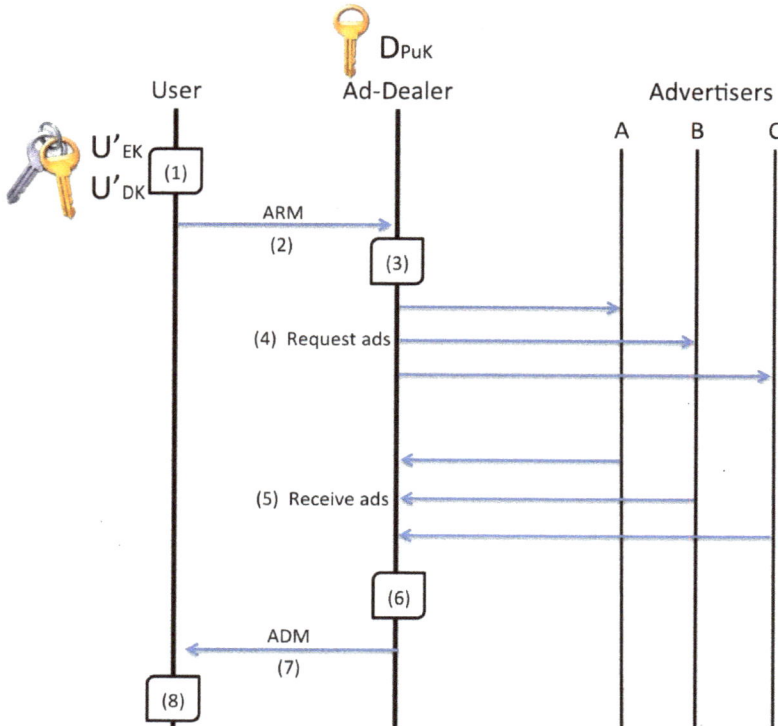

Figure 11. Advert Collection sub-protocol.

5.3.3. Ad Report Submission Sub-Protocol

The Advert Collection sub-protocol that is depicted in Figure 12 is used for the delivery of Advert Reports by the user to the Broker.

1. The user gradually composes his or her reports into an ARC as he or she is interacting with adverts. The contents of the ARC are encrypted with the Broker's public key B_{PuK} except for the first block that contains the $ARC\text{-}ID$ and the sequence number N and hash digest $h(N-1)$ in all remaining blocks. When the ARC is ready for submission, the user produces an Integrity Hash IH by computing $h(ARC)$ and signing the result with the signing key sk which were delivered to him with the sub-protocol described in Section 5.3.1.
2. The ARC and IH are sent to one of the Ad-Dealers either directly or via a Commissioner.
3. The Ad-Dealer keeps a local copy of the ARC and IH.
4. The Ad-Dealer submits the ARC and IH to the Broker.
5. The Broker first decrypts the ARC with his private key B_{PrK} and verifies the authenticity of the IH with the matching verification key vk. The Broker then verifies the integrity of the ARC by replicating the results of the hashes $h(N-1)$ in the individual blocks as well as the digest of $h(ARC)$ that is found in the IH. When verification has been completed successfully the Broker uploads the ARC and IH onto the $SC\text{-}Board$ in plaintext form. Finally, the Broker verifies the validity of the Checkpoint Blocks CBs and Affiliation Blocks ABs and Marks them on the $SC\text{-}Board$. The CBs are validated by checking the authenticity of the Ad-Dealers signature with a public key which is used only for this application. The ABs are validated by checking the signature (which exploits the sk' of the second user who participated in the meeting) and by cross referencing the dates in both the $ARCs$ (if the other ARC has already been submitted).
6. When the ARC and IH have been uploaded to the $SC\text{-}Board$, the Broker notifies the **submitting** Ad-Dealer with a Check Message.
7. The submitting Ad-Dealer verifies the hashes in the uploaded ARC and IH by comparing them to his own copy (recall that the hashes were not originally encrypted). The submitting Ad-Dealer then confirm the correctness of the ARC by placing his name and signature in the third field of the $SC\text{-}Board$.
8. The Broken notifies the Advertisers for the new entry in the $SC\text{-}Board$.
9. The Advertisers begin to reward the Publishers and Ad-Dealers and each reward report of the ARC is marked on the $SC\text{-}Board$ by the appropriate Advertiser. Each Advertiser is responsible for individually determining the honesty of the user who submitted the ARC by assessing the embedded authentication credentials which have been marked (CBs, ABs, $RoCs$ and $RoAs$). The $RoCs$ and $RoAs$ are validated and marked on the $SC\text{-}Board$ by the respective Advertisers after conforming the contained $C\text{-}Tokens$ and $A\text{-}Tokens$. Depending on the number and significance of credentials in the ARC, the Advertisers may choose to award a report or wait for more credentials to be marked on the $SC\text{-}Board$ (more awarded $RoCs$ and $RoAs$ by other Advertisers and more confirmed ABs).
10. After a certain time has passed from the submission of the ARC, the user's mobile client checks the $SC\text{-}Board$ in order to determine that his or her report has been registered. If the matching $ARC\text{-}ID$ is present within the $SC\text{-}Board$, the user may discard his or her original copy of the ARC and IH or else resubmits them. Recall that the only part of the $SC\text{-}Board$ which is visible to the user is the $ARC\text{-}ID$ while the rest is kept private among the remaining stakeholders.

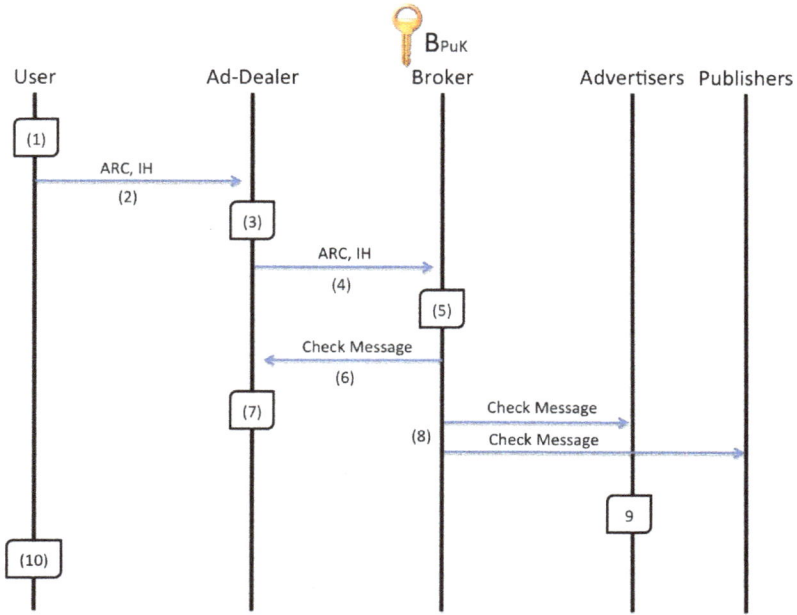

Figure 12. Ad Report Submission sub-protocol.

6. Evaluation

We follow a qualitative approach to evaluate our work by scrutinizing the performance of our system against the evaluation criteria that we established in Section 4.4. The goals are (1) to determine the effectiveness of our model at delivering accurate advert reports, (2) assess the system's resilience to fraudulent behaviour which is targeted against both the Advertisers as well as the Ad-Dealers and Publishers, and lastly (3) to evaluate the effectiveness of the system at preserving the privacy of the user.

6.1. Reporting Effectiveness

Reporting effectiveness is defined as the system's capability to deliver the necessary information which is required by the Publishers and Ad-Dealers (as a substitute of the Ad-Network) to claim their rewards from the Advertisers. As we do not process the data of Ad-Reports beyond encrypting and decrypting it, the information delivered by our system is the same as in the currently deployed system. In that regard, our system offers a similar reporting effectiveness in terms of the quality of information which it supports. In addition to this, our approach allows for the acquisition of supplementary information about the social habits of consumers. By matching the *ABs* (Affiliation Blocks) within the *ARCs* (Ad-Report Chain), the Broker would be able to deduce the consumer interest similarity of users that belong in the same social cycle. Furthermore, the Broker can also infer correlations between adverts, Publishers and locations (e.g., users who view adverts of product A are also interested in product B or users who visit location X tend to view adverts through Publisher Y). This offers a very useful insight into consumer practises which is obtained without violating the user's privacy by tracking his or her IP address.

As *ARCs* are shared on the *SC-Board* under specific *ARC-IDs*. This ensures that all stakeholders have access to them and that the same Ad-Report is not rewarded more than once. Moreover, the *SC-Board* allows partial access to users which enables them to detect when an *ARC* has not been delivered on time and may need to be resent, thus increasing the system's robustness. Although participation is voluntary, the privacy that the system provides is a good incentive for attracting users.

6.2. Fraud Protection of Advertisers

Financial fraud against the Advertisers is the main shortcoming of the currently enforced model that we address with our work. The main perpetrators of fraud are BotNets (automated clickers) and Click Farms where low-paid workers are hired to click on adverts. Such schemes commit fraud by generating a large bulk of Ad-Report traffic that does not correspond to actual consumer activity. To combat this problem, our model enables the affected stakeholders (Advertisers, Publishers and Broker) to (1) calculate the rate upon which a user creates reports and (2) verify that a user has the same behavioural patterns as a typical consumer.

To illustrate the system's protection against fraud, we will examine an attack scenario where a dishonest user $\hat{U}$ attempts to commit fraud at a large scale by submitting a set $\hat{S} = \{\widehat{ARC_1}, ..., \widehat{ARC_i}\}$ of fictitious *ARCs*. Recall that an *ARC* contains the following types of blocks: *RoV* (Report of View), *RoC* (Report of Click), *RoA* (Report of Action), *CB* (Checkpoint Block) and *AB* (Affiliation Block).

Among all types of reports, the *RoV* is the easiest to fabricate as it does not contain any verifiable information (*Tokens*). However, $\hat{U}$ cannot submit an $\widehat{ARC}$ which only contains *RoVs* as this would be immediately rejected by the Broker. The *RoC* and *RoA* contain a *C-Token* (Click-Token) or an *A-Token* (Action-Token) which can only be obtained by visiting the Advertiser's website within a particular time-frame. This prevents $\hat{U}$ from creating *RoCs* and *RoAs* ahead of time but $\hat{U}$ can still attempt to commit fraud by creating the $\widehat{ARC}$ over a longer period of time. Although this makes the creation of the $\widehat{ARC}$ more difficult, it is still possible with the use of an automated process that automatically downloads the *Tokens* when they become available. Despite this fact, the $\widehat{ARC}$ would still be rejected by the Broker as it would not contain any *CBs* or *ABs*.

Recall that *CBs* are distributed by Ad-Dealers and include a *T-Token* (Time-Token) which is signed and cannot be forged without knowledge of a secret cryptographic key. In order to obtain valid *CBs*, $\hat{U}$ would need to repeatedly travel to the physical location of an Ad-Dealer throughout the course of the creation possess of the $\widehat{ARC}$. Moreover, the fraudster must be cautious not to request multiple *CBs* (for different $\widehat{ARCs}$) at the same time as this would provoke suspicion. Even if $\hat{U}$ were to conspire with one of the Ad-Dealers, the $\widehat{ARC}$ would still be in danger of being exposed due to the disproportionate number of *CBs* from just one source. For such an attack to be successful, the fraudster would need to conspire with multiple Ad-Dealers and manage the *CBs* in a way that does not create an observable pattern (e.g., multiple $\widehat{ARCs}$ containing *CBs* from the same group of Ad-Dealers). Provided that the Ad-Dealers are carefully selected, such a scenario would be unlike.

The *ABs* are exchanged between users and serve a similar purpose as *CBs* as they can be used to determine the rate in which the reports were created. In contrast to *CBs* however, the *T-Token* which is contained in *ABs* is not signed by an Ad-Dealer but by another user. This makes *ABs* vulnerable to forgery as $\hat{U}$ can exchange *ABs* between multiple fake $\widehat{ARCs}$. However, if the fraudster was to compose a $\widehat{ARCs}$ in such a manner, suspicions would still be raised by the Broker due to the lack of *CBs*, *RoCs* and *RoAs*.

To conclude, in order for $\hat{U}$ to fabricate $\widehat{ARCs}$ which are realistic enough to fool the Broker, $\hat{U}$ would need to use an automated process which downloads *C-Tokens* and *A-Tokens* over an extended period of time. During that time, $\hat{U}$ would need to exchange *ABs* between the $\widehat{ARCs}$ and also physically collect *CBs* from different Ad-Dealers without raising their suspicion by submitting multiple requests at the same time. This process is time consuming and impractical which makes the conduct of financial fraud more difficult to accomplish.

6.3. Report Integrity

After an ARC leaves the user's device, it has to go through a Commissioner, an Ad-Dealer and the Broker before finally being posted on the SC-$Board$ (Service Confirmation Board). This makes it possible for any of the intermediaries to commit fraud by altering the content of an ARC. This type of fraud would be particularly difficult to detect due to the fact that a legitimate ARC (one created by a real user) is likely to have valid *Tokens*. Our system prevents this attack through the employment of hash functions and verification checks. To demonstrate the operation of the integrity mechanism, we will consider two attack scenarios.

Attack scenario 1:

The Commissioner and the submitting Ad-Dealer attempt to alter the content of a legitimate ARC in order to trick the Broker and Advertisers into rewarding a malicious Publisher for a publication that did not take place. Recall that the ARC follows the architecture of a blockchain where the first block holds a unique ARC-ID and each following block includes the hash digest of the previous. Additionally, the user also sends an IH (Integrity Hash) that contains the hash digest of the entire ARC which has been signed with a verification key vk.

Since the content of the Ad-Reports is encrypted with the Broker's public key B_{PuK}, it would be possible for a malicious Commissioner or Ad-Dealer to create a fictitious Ad-Report of his own. However, if the fictitious Ad-Report were to be inserted into the ARC (either as a new block or by replacing an existing one), this would result in a mismatch of both the hash digests within $ARC's$ blocks as well as the hash digest that is included in the IH. The attacker would be able to change the hashes in the ARC but cannot change the hash in the IH without knowing the user's verification key vk which was delivered to the user within the RF (Report Form). The RF was encrypted with a key that the user selected himself and only shared with the Broker via his public key B_{PuK}. This makes it impossible for an attacker to alter the content of IH without being exposed.

Attack scenario 2:

The Broker attempts to alter the content of a legitimate ARC in order to cheat a Publisher out of a reward. Although the Broker can be considered trustworthy as the representative of Advertisers, our system allows the Ad-Dealers and Publishers to verify that there has been no tampering of the Ad-Reports.

Recall that Ad-Reports are encrypted with the Broker's public key B_{PuK} and are shared on the SC-$Board$ (Service Confirmation Board) after they have been decrypted. The Broker could therefore attempt to cheat the Ad-Dealers and Publishers by altering the Ad-Reports of a submitted ARC before uploading it to the SC-$Board$. To prevent this attack, the submitting Ad-Dealer holds a copy of the ARC before forwarding it to the Broker. Although certain parts of the ARC are encrypted with Broker's public key B_{PuK}, the hash digests are transferred in plaintext. This allows the submitting Ad-Dealer to replicate the hash functions on the posted ARC in order to verify that decryption has been completed correctly. The submitting Ad-Dealer then marks the verification check VC-S in the SC-$Board$ which informs the Publishers and remaining Ad-Dealers that the submitted ARC is valid.

6.4. User Privacy

User privacy is ensured through the application of encryption, anonymous connections and the employment of Agents and Commissioners who serve the role of partially trusted proxies. The Advert Request Message (ARM) which is sent by the user to request adverts is encrypted with the Broker's public key B_{PuK}. The requested adverts are placed in an Advert Delivery Message (ADM) that is encrypted with an encryption key U'_{EK} which was provided by the user within the ARM. This ensures that the Agent who serves the user has no access to neither the requests nor the ads. Similarly, the Ad-Reports of the user's ARC are also encrypted with the Broker's public key B_{PuK} which ensures privacy against the serving Commissioner. The Broker does not learn the identities of the user, the Agent or the Commissioner as the exchange of information takes place over an anonymous connection. The same anonymous connection is also in place when the user collects a Checkpoint Block (*CB*) from an Ad-Dealer within his or her proximity. The request is encrypted with the Ad-Dealer's public key D_{PuK} and contains a user-generated encryption key U_{EK} which is used by the Ad-Dealer to transfer the *CB*. This ensures that an eavesdropper cannot gain access to the user's information nor spoof the identity of the Ad-Dealer in order to transmit a fake *CB*.

7. Conclusions

Digital advertising allows Advertisers to effectively target users with promotional materials and at the same time offers a sustainable source of income for Publishers. Advertising fraud however, presents a serious threat for the marketing industry as it is costing Advertisers billions of dollars every year. The attempts of Ad-Networks to contain the problem by filtering incoming traffic has had limited effectiveness against BotNets and Click-farms who employ various methods in order to remain undetected.

In this paper, we address the problem by contributing an alternative reporting method which filters Ad-Reports by determining if they were composed by honest or dishonest users. What constitutes a user's honesty is the rate at which they create Ad-Reports. While honest users view adverts at a paced manner over a longer period, dishonest users (whether BotNets or Click-farms) generate large volumes of reports in a short amount of time. The challenge that we face is the need to protect the user's anonymity which prevented us from simply verifying real users through traditional means such as digital signatures. To overcome this limitation, our approach identifies honest users based on their behavioural patterns. Typical social behaviours such as making online purchases, traveling to specific locations and engaging with other users are exploited by our system in order to allow users to collect a series of digital credentials which we call *Tokens*. The *Tokens* work as proof that the user who collected them was engaged in a specific social activity (e.g., traveling to designated location) at a particular time. When combined with Ad-Reports in the form of an *ARC* (Ad-Report Chain) which follows the same architecture as that of a blockchain, the *Tokens* operate both as time-stamps which give us the ability to determine the rate the reports were created at but also offer proof that the creator has the type of social activity which is indicative of an honest user. The *Tokens* do not include any identifying information and the *ARC* is then delivered through the use of opportunistic networking and anonymous connections which ensure the user's privacy. This enables Advertisers to validate multiple reports which are composed by the same user without the user's identity being exposed. The combination of effective Ad-Report verification and user anonymity offers a well-balanced solution where both the Advertisers and users achieve their goals without one of them needing to make a compromise. Our method of verifying users based on their behaviour creates a promising alternative to verification methods which involve tracking and therefore violate the user's privacy.

Author Contributions: Stylianos S. Mamais, System design and manuscript creation; George Theodorakopoulos, Corrections and suggestions.

Conflicts of Interest: The authors declare no conflict of interest.

References

1. Digital Ad Spending to Surpass TV Next Year-eMarketer. 2016. Available online: https://www.emarketer.com/Article/Digital-Ad-Spending-Surpass-TV-Next-Year/1013671 (Accessed on 10 May 2017).
2. What Is an Untrustworthy Supply Chain Costing the US Digital Advertising Industry? 2015. Available online: http://www.iab.com/wp-content/uploads/2015/11/IAB_EY_Report.pdf (Accessed on 11 May 2017).
3. Mungamuru, B.; Weis, S.; Garcia-Molina, H. *Should ad Networks Bother Fighting Click Fraud? (Yes, They Should)*; Technical Report; Stanford InfoLab Publication Server, Stanford, CA, USA , 2008.

4.   Vidros, S.; Kolias, C.; Kambourakis, G.; Akoglu, L. Automatic detection of online recruitment frauds: Characteristics, methods, and a public dataset. *Future Internet* **2017**, *9*, 6.

5.   Walgampaya, C.; Kantardzic, M.; Yampolskiy, R. Real time click fraud prevention using multi-level data fusion. In Proceedings of the World Congress on Engineering and Computer Science, San Francisco, CA, USA, 20–22 October 2010; Volume 1, pp. 20–22.

6.   Juels, A.; Stamm, S.; Jakobsson, M. Combating click fraud via premium clicks. In Proceedings of 16th USENIX Security Symposium on USENIX Security Symposium, Boston, MA, USA, 6–10 August 2007; USENIX Association: Berkeley, CA, USA, 2007; pp. 2:1–2:10.

7.   Haddadi, H. Fighting online click-fraud using bluff ads. *ACM SIGCOMM Comput. Commun. Rev.* **2010**, *40*, 21–25.

8.   Iqbal, M.S.; Zulkernine, M.; Jaafar, F.; Gu, Y. Fcfraud: Fighting click-fraud from the user side. In Proceedings of the 2016 IEEE 17th International Symposium on High Assurance Systems Engineering (HASE), Orlando, FL, USA, 7–9 January 2016; pp. 157–164.

9.   Faou, M.; Lemay, A.; Décary-Hétu, D.; Calvet, J.; Labrèche, F.; Jean, M.; Dupont, B.; Fernande, J.M. Follow the traffic: Stopping click fraud by disrupting the value chain. In Proceedings of the 2016 IEEE 14th Annual Conference on Privacy, Security and Trust (PST), Auckland, New Zealand, 12–14 December 2016; pp. 464–476.

10.   Pizzolante, R.; Carpentieri, B.; Castiglione, A.; Castiglione, A.; Palmieri, F. Text compression and encryption through smart devices for mobile communication. In Proceedings of the 2013 IEEE Seventh International Conference on Innovative Mobile and Internet Services in Ubiquitous Computing (IMIS), Taichung, Taiwan, 3–5 July 2013; pp. 672–677.

11.   Papadimitratos, P.; Haas, Z.J. Secure routing for mobile ad hoc networks. In Proceedings of the SCS Communication Networks and Distributed Systems Modeling and Simulation Conference (CNDS), San Antonio, TX, USA, 27–31 January 2002; pp. 193–204.

12.   Arapinis, M.; Mancini, L.I.; Ritter, E.; Ryan, M.D. Analysis of privacy in mobile telephony systems. *Int. J. Inf. Secur.* **2017**, *16*, 491–523.

13.   Traynor, P.; Lin, M.; Ongtang, M.; Rao, V.; Jaeger, T.; McDaniel, P.; La Porta, T. On cellular botnets: Measuring the impact of malicious devices on a cellular network core. In Proceedings of the ACM 16th ACM Conference on Computer and Communications Security, Chicago, IL, USA, 9–13 November 2009; pp. 223–234.

14.   Lilien, L.; Kamal, Z.H.; Bhuse, V.; Gupta, A. Opportunistic networks: The concept and research challenges in privacy and security. In Proceedings of the WSPWN, Miami, FL, USA, March, 2006; pp. 134–147.

15.   Boldrini, C.; Conti, M.; Passarella, A. Exploiting users' social relations to forward data in opportunistic networks: The HiBOp solution. *Pervasive Mob. Comput.* **2008**, *4*, 633–657.

16.   Theodorakopoulos, G.; Boudec, J.Y.L.; Baras, J.S. Selfish response to epidemic propagation. *IEEE Trans. Autom. Control* **2013**, *58*, 363–376.

17.   Mamais, S.S.; Theodorakopoulos, G. Private and secure distribution of targeted advertisements to mobile phones. *Future Internet* **2017**, *9*, 16.

18.   Fain, D.C.; Pedersen, J.O. Sponsored search: A brief history. *Bull. Am. Soc. Inf. Sci. Technol.* **2006**, *32*, 12–13.

19.   Narayanan, A.; Thiagarajan, N.; Lakhani, M.; Hamburg, M.; Boneh, D. *Location Privacy via Private Proximity Testing*; NDSS; Stanford University: Stanford, CA, USA, 2011; Volume 11.

future internet

MDPI

Article

Private and Secure Distribution of Targeted Advertisements to Mobile Phones

Stylianos S. Mamais and George Theodorakopoulos *

School of Computer Science and Informatics, Cardiff University, 5 The Parade, Roath, Cardiff CF24 3AA, UK;
MamaisSS@cardiff.ac.uk
* Correspondence: TheodorakopoulosG@cardiff.ac.uk Tel.: +44-29-208-74855

Academic Editor: Georgios Kambourakis
Received: 4 April 2017; Accepted: 26 April 2017; Published: 1 May 2017

Abstract: Online Behavioural Advertising (OBA) enables promotion companies to effectively target users with ads that best satisfy their purchasing needs. This is highly beneficial for both vendors and publishers who are the owners of the advertising platforms, such as websites and app developers, but at the same time creates a serious privacy threat for users who expose their consumer interests. In this paper, we categorize the available ad-distribution methods and identify their limitations in terms of security, privacy, targeting effectiveness and practicality. We contribute our own system, which utilizes opportunistic networking in order to distribute targeted adverts within a social network. We improve upon previous work by eliminating the need for trust among the users (network nodes) while at the same time achieving low memory and bandwidth overhead, which are inherent problems of many opportunistic networks. Our protocol accomplishes this by identifying similarities between the consumer interests of users and then allows them to share access to the same adverts, which need to be downloaded only once. Although the same ads may be viewed by multiple users, privacy is preserved as the users do not learn each other's advertising interests. An additional contribution is that malicious users cannot alter the ads in order to spread malicious content, and also, they cannot launch impersonation attacks.

Keywords: online behavioural advertising; opportunistic networks; mobile networks; privacy; privacy-preserving advertising

1. Introduction

Advertising has played a very significant role in the expansion of the digital media industry over the past few years. The ability to generate revenue through the publishing of adverts allows companies to offer a wide range of services without any cost for users while still making profit.

Although digital advertising has been around for almost as long as the Internet itself, the latest financial studies suggest that the advertising market has experienced a rapid increase in recent years with a prime focus given to adverts that are specifically designed to target mobile devices. This shift towards mobile ads is driven by feedback from the online browsing habits of the consumers themselves [1]. Research analysis indicates that on average, an adult will spend more than 2.5 h a day on his/her smartphone [1]. In the U.K. alone, spending on mobile ads has increased by 45% in the year 2015. This accounts for an expenditure of 3.2 billion pounds, which is predicted to further increase by another 35% in the following years [2]. With such investments on mobile advertising, it should come as no surprise that a great deal of effort is focused towards effective advert targeting.

Online Behavioural Advertising (OBA) is a highly effective method for distributing content-aware adverts that target consumers based on their individual interests [3]. From a business point of view, OBA is beneficial to both vendors and consumers as the former increase revenue and the latter are only presented with ads that are relevant to their needs. Regardless of the commercial benefits, targeted advertising raises serious privacy concerns.

1.1. Problem Statement

Targeting of content-aware adverts is only possible after analysing a user's behavioural conducts. This practice utilizes private data that may reveal sensitive information, such as demographics, ideological and religious inclinations, shopping and browsing practices, lifestyle preferences and pretty much any other intelligence that can be deduced depending on the platform.

Although OBA has been employed for years, the ever-increasing usage of smartphones further escalates the existing problem. In addition to being more pervasive than computers, mobile devices incorporate a multitude of tracking capabilities that computers simply do not have. Smartphones have access to information such as personal text messages, emails, GPS coordinates, WiFi access point locations, purchasing habits through applications, such as Apple pay or loyalty point apps, and activity sensors.

Although the general public may not be fully aware of the technical details of OBA, numerous relevant studies indicate that the majority of people disapprove of being tracked. Specifically, a survey conducted by Pew Research Center showed that 68% of Americans view targeted ads in a negative way [4]. Despite more than half of the population disapproving of ads, we still need to acknowledge that targeted ads are vital for the offering of free services. Finding an alternative method of delivering personalized ads without compromising privacy is therefore an important problem that deserves to be addressed.

1.2. Contribution and Paper Layout

In this article, we aim to address the privacy concerns of targeted advertising by offering an alternative ad-distribution system. The contribution of our system compared to existing approaches is that it eliminates the need for trust among users, reduces memory and bandwidth overhead and also addresses the issue of security against impersonation and fake-ad injection attacks. We begin by illustrating the operation of the current OBA model and indicating its issues in Section 2. In Section 3, we examine and classify other proposed ad-distribution systems and then evaluate them in terms of privacy, security, targeting effectiveness and practicality. In Section 4, we first consider our observations from the evaluation of the relevant work and then determine a threat model, as well as a set of operational requirements, which will serve as evaluation criteria for our system, which we present in Section 5. In Section 6, we evaluate our system with respect to the relevant work by examining different threat scenarios, which are determined based on the threat model and evaluation criteria that we define in the previous section. Finally, in Section 7, we summarize our findings and present our conclusions.

2. The OBA Model

Online Behavioural Advertising (OBA) is defined by the United States Federal Trade Commission (FTC) as "the practice of tracking an individual's online activities in order to deliver advertising that is tailored to the individual's interests" [5].

The OBA system that is currently operational consists of four components. The central component is known as the Ad-Networkand is responsible for monitoring the users' behaviour in order to present them with relevant ads. Directly connected to the Ad-Network are the Advertisers and the Publishers. Advertisers operate as representatives of businesses who wish to promote products and services through advertising campaigns. The Advertisers create promotional material and supply it to the Ad-Network along with pricing details, i.e., along with the price they are willing to pay to the Ad-Network every time their ads are being clicked on or viewed. The Ad-Network is responsible for delivering the ads to the consumers.

The adverts reach consumers by being displayed on platforms that are commonly viewed by users. These platforms are known as Publishers and are primarily websites, software applications or other popular services that are regularly visited by users. For an agreed price, Publishers give full control of certain visual areas within their websites or software to the Ad-Network. These areas are known as Ad-Boxes and they are used to display adverts. When a user visits a website or service of a Publisher, the Ad-Network chooses one of the adverts that have been supplied by the Advertisers and then displays it within the Ad-Box.

3. Related Work

Increasing the privacy in targeted advert delivery models is a research topic that several academics have attempted to resolve. Researchers seem to agree that sensitive user data should be kept outside the reach of the Ad-Network, Advertisers, Publishers and any other party that is not considered trusted.

In the following sections, we offer insight into the available advert delivery systems. We first classify them into three categories based on the approaches and methods that they use, and then, we provide an overview for some of the most notable systems for each category. Finally, we conduct a critical evaluation for each category and proceed to recognize their flaws and limitations.

3.1. Classification of Available Systems

Previously-proposed advert delivery systems incorporate various combinations of architectures, as well as privacy mechanisms and can be arranged in three general categories.

- Trusted proxy-based anonymity.
- Selection from pool of adverts.
- Anonymous direct download.

3.1.1. Trusted Proxy-Based Anonymity

The simplest method of achieving anonymity is by introducing some form of trusted third party that acts as a proxy between the user and Ad-Network. The role of the proxy is to mask the identity of the user by forwarding his/her requests after replacing any identifying information with a temporary identifier. The Ad-Network uses this temporary identifier to reply to the proxy with relevant adverts, which are then conveyed back to the user.

In order to further increase privacy, public key encryption can be used to encrypt requests and adverts. When paired with cryptography, the proxy is aware of the identity of the user that is sending the requests, but cannot see the content of the requests nor the corresponding ads. In turn, the Ad-Network or any other entity that distributes ads can decrypt the encrypted requests, but is not aware of the user's true identity, as it is masked by the proxy.

3.1.2. Selection from Pool of Adverts

Schemes that are based on this approach make use of client-side processing by allowing users to select ads that best satisfy their advertising needs, out of a pool of ads. The pool can be populated by various methods with the simplest one being by making a generic request. When following this approach, the user issues a request for ads that fall under a very broad category, which includes his/her specific interest. For example, if the user is interested in running footwear, he/she may make a generic request for sporting equipment. The Ad-Network responds with multiple ads that satisfy the request, and it is up to the user to keep ads that best match his/her particular interest and discard the rest.

3.1.3. Anonymous Direct Download

A further group of proposed approaches directly download freely broadcast adverts through the use of specialized hardware and software. Advertisers store their ads at broadcasting stations that operate in publicly-accessible locations. As a user comes into proximity of these stations, his/her device downloads the available adverts. The user's device is then responsible for sorting through the collected ads and selecting the most relevant, while the rest are discarded.

Anonymity is achieved through the use of protocols that enable mobile devices to connect to the broadcasting stations without disclosing of any information that exposes the user's identity, such as username, network address or physical address. Some of these systems also enable the users to connect and exchange adverts with each other. In these systems, a user downloads ads from a broadcasting station and then he/she propagates them to other users that he/she later comes into proximity with. This extends the reach of the broadcasting stations, but it also requires a certain level of trust among the users.

3.2. Literature Review

The P2PMM-system [6] relies on a trusted proxy that is referred to as the Intermediary Services Provider (ISP). The ISP is entrusted to store the sensitive information of the user and directly answer requests with adverts that it has obtained from merchants. Although the ISP has no immediate interest to expose any information to the merchants, this method assumes that the ISP can be fully trusted. Obliviad [7] accomplishes a similar goal by replacing the proxy with a hardware device that is placed on the Ad-Network side. The device receives the requests from the client and then sends a number of matching ads that are obtained from the Ad-Network's database. The system accounts for click reports and maintains privacy by deploying a Private Information Retrieval (PIR) mechanism, which allows the client to access the Ad-Network's database, while preventing the Ad-Network from learning about the query and the resulting answer [8]. Aside from requiring additional computational power, this architecture does not guarantee that the operator of the Ad-Network will not bypass security by physically tampering with the device.

Adnostic [9] composes a local interest profile that does not get disclosed to other parties. When the user visits a website, he/she is sent a number of ads that are relevant to the contextual theme of that particular website. For example, if the user visits a travel website, the ads can be for holidays to various destinations. The ad that is the most relevant according to the user's interest profile is then selected and displayed, while the rest are discarded. Although this method may be considered secure, it produces unnecessary overhead. It can also be argued that targeting is not very effective, since the ads that are sent to the user are based on a very general interest assumption. Kodialam et al. [10] follow a similar approach as [9]. They propose a role reversal scheme where the ad providers send to the user a series of interest profiles along with a set of matching adverts. This approach may potentially be more effective than Adnostic [9], since the users have a wider variety of ads to choose from. However, the generated overhead due to unused ads still degrades efficiency. Privad [11] is based on the selection from an advert pool, but also incorporates a proxy. A trusted third party, the Dealer, operates between the user and the Ad-Network. The user selects a general interest category and sends it to the Ad-Network through the Dealer (proxy). Upon receiving the message, the Ad-Network uses the same path to respond with a wide variety of relevant ads. The user's device sorts through these ads and selects the most relevant to be displayed. Although this method is simple to incorporate into the existing model, it assumes the existence of a fully-trusted third party. The Dealer is also a single point of failure, and if compromised, the security of the entire system can be bypassed.

PervAd [12,13] provides personalized ads through broadcasting. Users who maintain a local interest profile can collect relevant ads as they move into proximity of customized WiFi access points. The system minimizes overhead by first sharing some contextual information about the available ads, thus allowing users to selectively download only specific content. The interest profile is specified by the user himself/herself, and the downloading process is performed anonymously. Even though this method achieves a substantial level of privacy, it is highly impractical as users need to physically travel to specific locations. The schemes in [14,15] also use broadcasting, but at the same time make use of opportunistic networking. In both schemes, the users download adverts directly from local businesses. As users come into proximity, their devices connect and exchange ads based on contextual information.

A mechanism for keeping track and rewarding points for interactions is also established. Although this is necessary for identifying the users who contribute the most to the system by propagating ads to others, it may also be a threat to the privacy of the system as it exposes the identity of the users, as well as their social encounters. An additional factor that the system does not account for is the presence of malicious users that may affect the integrity of the system by spreading fake ads or malware.

The Let's Meet! [16] framework uses a client-server architecture which establishes a cooperation link between mobile users who share an interest for a particular offer, but may be unrelated to each other. More specifically, Let's Meet! enables consumers to take advantage of group offers by physically bringing them together in the location of a local vendor. The authors emphasize privacy and security by incorporating mechanisms that prevent the disclosure of sensitive consumer information and defend against malicious users who may launch impersonation attacks or attempt to forge offer-coupons.

3.3. Classification Assessment

Our assessment of the related work is shown in Table 1. The previously-proposed systems still have certain shortcomings, which we will now analyse in more detail.

Systems that make use of a proxy can be easily incorporated into the current architecture and achieve an adequate level of privacy, without seriously reducing the effectiveness and the efficiency of the system. Regardless, such models assume the existence of a trusted or partially trusted third party that can act as the proxy. This is not entirely realistic and also creates a single point of failure that threatens the integrity of the system if compromised.

Systems that are based on the selection from a pool of adverts, maintain privacy by taking advantage of the computational capability of mobile devices. These systems do not require any trust between the participants, and depending on the method that is used to populate the pool, they can achieve a satisfactory level of privacy. Although these systems are not too difficult to introduce, they suffer greatly in terms of targeting effectiveness and resource efficiency. The information that is shared with the Ad-Network may be too generic to effectively retrieve adverts that perfectly correspond to the user's interests. A significant amount of overhead is also generated. This is due to the fact that the advert selection needs to be performed locally and also additional ads need to be downloaded and stored only to be discarded afterwards.

Table 1. Assessment of categories of ad-delivery systems (good: "+"; adequate: "×"; poor: "−").

	Trusted Proxy	Pool of Ads	Direct Download
Privacy vs. Ad-Network	×	×	+
Privacy vs. other users			−
Security vs. attacker	+	+	−
Targeting effectiveness	+	−	+
Practicality/usability	+	−	−
Resource conservation	+	−	×

Models that allow users to directly download content from broadcasting stations offer the highest level of security so far. In some of these systems, the user does not even need to share any information, and therefore, privacy is only dependent on the capability of mobile devices to connect to stations anonymously. These systems require the use of specialized equipment that may be costly or impractical to incorporate. Their effectiveness to deliver accurate content depends on the amount of information that the user shares with the station, the availability of relevant adverts and the user's ability to physically travel to the specified locations. Resource efficiency is also limited, as these models can be very demanding in terms of storage, processing power and battery life. When enhanced with opportunistic networking, the effectiveness of the system may increase, but this will also raise trust issues among the users. Users need strong incentives to actively participate in such systems and efficiency will be reduced even further as additional strain will be put on the mobile devices.

4. Proposed System Overview

The system that we propose is based on the traditional anonymous download architecture and also incorporates opportunistic networking. Fine-grained targeted adverts are dispatched from broadcasting stations and then are propagated across the nodes (users) of a social network. Our system improves on previous attempts, as it eliminates the need for trust among users, reduces memory and bandwidth overhead and also addresses the issue of security against impersonation and fake-ad injection attacks.

Our system is designed to be highly versatile in terms of ad-targeting and opportunistic routing. The targeting algorithm, which is the mechanism that determines the user's advertising needs, can operate independently of the rest of the system for as long as it produces an output of a standard format. The opportunistic routing algorithm that we use in this paper follows a probabilistic approach, which is inspired by PRoPHET [17]. However, our model is also compatible with a wide range of other opportunistic networking schemes.

4.1. System Stakeholders

Our system consists of four basic stakeholders that operate independently from each other and in accordance with their own incentives and goals. Advertisers are advertising firms that wish to promote products through advertising much like in the OBA system. Advertisers commission the services of a Broker, which is a privately owned company that generates revenue by distributing ads to consumers. The Broker delivers the adverts through a network of distributors that are called Ad-Dealers. Ad-Dealers have a physical presence in publicly accessible areas and are able to deliver adverts upon request by establishing wireless connections with the mobile devices of users. Users represent potential consumers who own smartphone devices and are targeted by advertising companies through the services they use, such as websites and free mobile apps.

The Broker is responsible for the Ad-Dealers in his/her network and therefore needs to institute some form of trusted relationship by performing background checks and signing legal agreements. Ideal to play the role of the Ad-Dealers are local entities that have the capability to install and manage advert distribution hardware in publicly-accessible locations such as malls, shops and local WiFi hotspots. The Ad-Dealers conduct their own business independently but at the same time share resources and work cooperatively under the administration of the Broker.

All of the participants operate independently and the system can easily be expanded. The Broker can recruit new Ad-Dealers by authorizing the installation and use of the required infrastructure, while new users only need to download and register the client software on their devices.

4.2. Trust and Threat Model

Advertisers are assumed to fully trust the Broker to distribute their adverts, as well as to provide them with accurate billing information. We need to point out that this assumption is not entirely realistic as the Broker has a great interest to lie to the Advertisers in order to overcharge them. As the main focus of this research is the privacy of the user, this constitutes an entirely different problem that only concerns the Advertisers and the Brokers. This problem is also present in the current OBA model and can be addressed in the future in a manner that does not affect the rest of the system.

The Ad-Dealers are considered by the Broker as trustworthy enough not to compromise the integrity of the system by exposing secret cryptographic keys or tampering with the infrastructure. No trust is required on behalf of the Ad-Dealers towards the Broker as he/she has no means or interest to interfere with the Ad-Dealers' business.

The Ad-Dealers, as well as the Broker are considered to be honest enough to provide relevant adverts that are not fake or malicious, but at the same time they are curious and very determined to obtain access to private data. The user can therefore trust the provided material but is not willing to expose any information that can link his/her true identity (name, address, banking information) to any specific advertising preferences.

Despite being part of the same social circle, users do not fully trust each other. A user can easily be compromised and act maliciously by revealing private information or by propagating content that is fake and harmful. Compromised users are also considered as a threat by the Broker and Ad-Dealers, as they can cause downtime or undermine the quality of service by attacking the system.

4.3. Evaluation Criteria

In consonance with the security threats that we recognize in the previous section and the shortcoming of analogous systems, we can now define a set of requirements that can serve as evaluation criteria for our system.

- User privacy against the Broker and Ad-Dealers: Neither the Broker nor his/her Ad-Dealers should be able to obtain any information that could be used to link a user's true identity to his/her advertising interests.
- User privacy against other users: Users should not have precise knowledge of the advertising interests of other users within their social circle.
- Protection from fake or harmful content: Attackers should not be able to infect the system with adverts that have not been distributed from a valid Ad-Dealer.
- Protection against impersonation attacks: Attackers should not be able to impersonate the identity of a user or an Ad-Dealer.
- Resource conservation: As mobile devices offer limited resources, the system needs to be conservative in the consumption of memory and battery power.

5. Detailed System Description

The Broker initiates the operation of the system by creating a grid of Ad-Dealers who are given access to specialized networking equipment and two public/private key pairs, one used for encryption/decryption and one for authentication. The Broker then develops the mobile client software and makes it accessible for users to download. Finally, the Broker accumulates ads from various Advertisers and forwards them to the Ad-Dealers. The ads are organized into groups based on their targeted audiences. When appearing at close proximity of an Ad-Dealer, the mobile client on a user's device can establish an anonymous connection and download ads for specific consumer interests on request. Additionally, users forward requests and adverts among themselves. When two mobile clients (users) A and B come into proximity, A sends ad requests to B. User B is responsible for collecting the relevant adverts from the Ad-Dealer by forwarding the requests and then delivering the ads back to A. The system comprises a sequence of sub-protocols that are triggered when certain events happen, e.g., when two users meet. We describe these sub-protocols and the corresponding triggers in Figure 1.

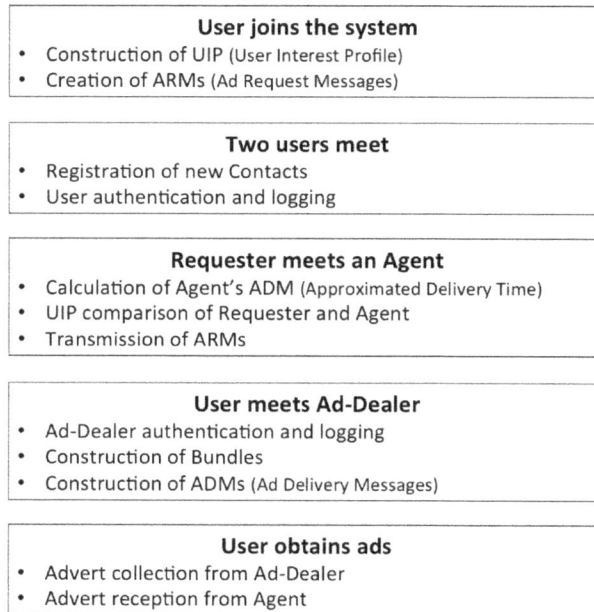

| **User joins the system** |
| • Construction of UIP (User Interest Profile) |
| • Creation of ARMs (Ad Request Messages) |

| **Two users meet** |
| • Registration of new Contacts |
| • User authentication and logging |

| **Requester meets an Agent** |
| • Calculation of Agent's ADM (Approximated Delivery Time) |
| • UIP comparison of Requester and Agent |
| • Transmission of ARMs |

| **User meets Ad-Dealer** |
| • Ad-Dealer authentication and logging |
| • Construction of Bundles |
| • Construction of ADMs (Ad Delivery Messages) |

| **User obtains ads** |
| • Advert collection from Ad-Dealer |
| • Advert reception from Agent |

Figure 1. System sub-protocols and their corresponding triggers.

5.1. User Joins the System

5.1.1. Construction of UIP

Smartphones have access to a multitude of user information such as browsing logs, emails, text messages, mobile app data, GPS coordinates and data from available sensors, such as accelerometers, pedometers and heart rate monitors. By simply tapping into these resources, the client can determine the users advertising needs with much higher effectiveness than a remote observer. For example, when the device detects high physical activity while in close proximity to a park, this can be associated to an advertising interest for sporting equipment.

The user's advertising interests are represented in the form of a list that follows a standard format throughout the entire system and is termed the User's Interest Profile (UIP). Each entry of the UIP denotes a common consumer interest (automotive, technology, food and drink, etc.) and is recognized by a unique identifier, which is noted as the Interest-ID. When given the right permissions, the client associates the user's activity with specific advertising interests that can in turn be marked on the UIP in a simple binary manner where "true" represents that the user's activity matches the particular interest of the UIP and "false" that it does not. The UIP is stored locally where it can be dynamically maintained in accordance to the alterations of the users behaviour and is not accessible to anyone else.

The determination of suitable interests happens independently from the rest of the system, thus offering a great deal of versatility. Developers will be able to fine-tune the system by designing their own algorithms that can be fully compatible with the rest of the system for as long as they produce an output which follows a standard UIP format.

5.1.2. Creation of ARMs

Once the UIP has been constructed, the marked interests that have been associated with the user's activity may be used to create requests for relevant adverts while the remaining entries are ignored. The client begins by generating a cryptographic key K_0^{user}, which is then used to create multiple one-time keys K_i^{user}, which are produced with the use of a hash chain:

$$K_i^{user} = h(K_{i-1}^{user}) \tag{1}$$

The client goes through the UIP and recovers the Interest-ID for each of the marked entries. For each of the marked interests, the client then creates an Ad Request Message (ARM), which contains the equivalent Interest-ID and a new K_i^{user}. The ARMs are then encrypted with a public key that comes pre-installed in the client and is known as the System Encryption Key (EK_{System}). Finally, the resulting encrypted ARMs, as well as a copy of each K_i^{user} are stored for future use. Note that this operation does not need to take place in real time. We can therefore significantly preserve resources by performing it while the device is idle and preferably connected to a power source.

$$ARM_{(ARM-ID)} = E_{EK_{System}}\left[interest - ID, K_i^{user}\right] \tag{2}$$

5.2. Two Users Meet

Mobile clients perform periodic scans (e.g., via Bluetooth or WiFi) to detect nearby Ad-Dealers and other clients. We need to note that in practice, there might be occasions where rapid encounters will be detected as two devices come in and out of range. For this reason, two encounters are considered as separate events only when a certain amount of time passes in between.

5.2.1. Registration of New Contacts

When two mobile devices come into proximity for the first time, both clients request a manual confirmation from the users so that they can register each other as Contacts. Registration happens by trading a unique identity number that is built within each version of the client and will from now on be simply referred to as User-ID. An out-of-band channel (SMS, QR code, email or keyboard input) is then used to exchange a pair of authentication passwords, and the meeting is logged. Alternatively, the passwords could be exchanged automatically via Bluetooth. Although this would simplify the operation, it would also create the risk of passwords being sniffed by a nearby eavesdropper. The passwords are generated automatically by a secure random string generation function (more details on this are given in Section 6.

5.2.2. User Authentication and Logging

When two registered Contacts meet, their encounter is logged by both clients after the users have authenticated each other with a challenge-response password handshake.

As shown in Figure 2, the procedure begins with the two users A and B exchanging their User-IDs (U_A and U_B) and two random nonce challenges R1 and R2. User A hashes R2 concatenated with his/her password P_A and sends the result $H_A = h(P_A, R2)$ to User B. Similarly, User B produces $H_B = h(P_B, R1)$ and sends it to A.

Once H_A and H_B have been delivered, User A generates $H_B' = h(P_B', R1)$ where P_B' is his/her originally stored copy of the password that B sent when they became Contacts, while User B performs the same operation to calculate $H_A' = h(P_A', R2)$. The two users can now authenticate each other by simply comparing the hash digests H_A with H_A' and H_B with H_B'.

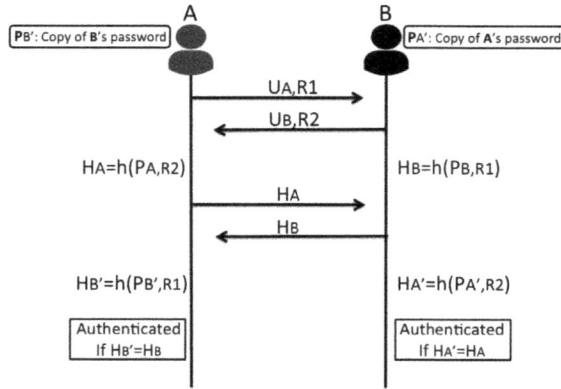

Figure 2. User handshake protocol for user authentication.

5.3. Requester Meets an Agent

As we explain in Section 4, adverts are propagated across the system through the use of opportunistic networking. Users download ads for their own use, but at the same time can collect ads on behalf of other users. This is highly beneficial for users that do not encounter Ad-Dealers often: such users can acquire adverts by exploiting the mobility of other users within the same social circle such as family members, friends and co-workers.

A user that ferries requests or adverts on behalf of another will from now on be called the Agent while the user who places the request will be called the Requester. The system can support Agents who ferry ads for multiple Requesters, and it can also support users who operate as both Requester and Agent (for another Requester) at the same time. However, for simplicity, we will examine the most basic scenario where two users meet and one of them plays the role of the Requester while the other serves as his/her Agent.

5.3.1. Calculation of Agent's DT

The opportunist network that we use operates as follows. A Requester sends ARMs to an Agent who forwards them to the Ad-Dealer when the two come within range. The Agent then receives from the Ad-Dealer ads that he/she conveys back to the Requester the next time they meet. For this scheme to be efficient, it is important that the Requester only uses the services of Agents who can deliver ads quicker than he/she would normally need to collect by himself/herself by visiting the Ad-Dealer.

The system adopts a history-based approach to determine the probability of the Requester receiving ads quicker if he/she were to use an Agent. This approach uses three metrics. The first metric is named Time To Ad-Dealer of a user (TTA_{user}), and it represents the approximate delay until the user is expected to appear again within the range of an Ad-Dealer. The TTA_{user} can be estimated by calculating the average time between successive encounters of the user with and the Ad-Dealers and subtracting from it the time that has passed since their last meeting (e.g., if the user encounters an Ad-Dealer every 24 h on average and it has been 4 h since the last encounter, then the TTA_{user} is $TTA_{user} = 24 - 4 = 20$ h).

The second metric is the Encounter Average $EA_{(u1,u2)}$ of two users, $u1$ and $u2$, and it denotes the average meeting time between successive encounters for a pair of users. It is obvious that since both users record the same meetings, $EA_{(u1,u2)}$ is expected to be equal to $EA_{(u2,u1)}$.

These two first metrics are calculated by both the Requester and the Agent with information that can be found in their own log files. The last metric is the Delivery Time of Agent (DT_{Agent}) and expresses the estimated period needed for a particular Agent to deliver an ARM to an Ad-Dealer and then convey back the requested advert. The DT_{Agent} is calculated only by the Requester, but also requires input from the Agent. The Requester initiates the operation by sending a message to express his/her intent to request ads. The Agent who receives the message first checks his/her availability by looking at different parameters such as his/her free buffer space and then responds either with a negative or positive reply that is also accompanied with his/her TTA_{Agent} and the total number of requests that he/she is willing to take.

Upon receiving the TTA_{Agent}, the Requester can use it in combination with the $EA_{(Requester, Agent)}$ ($EA_{(R,A)}$ for short) that he/she already possesses, to calculate an DT_{Agent}. The DT_{Agent} can be calculated as follows: if the current time is $T_0 = 0$, the Requester can assume that his/her future encounters with the Agent are expected to occur between intervals of time that are equal to $EA_{(R,A)}$. We can therefore approximate the time of expected future encounters as $T_1 = T_0 + EA_{(R,A)}, T_2 = T_1 + EA_{(R,A)}, ..., T_n = T_{(n-1)} + EA_{(R,A)}$.

Delivery will occur upon one of these encounters and more specifically upon the encounter that will take place right after the Agent has visited the Ad-Dealer. The DT_{Agent} is therefore equal to $T_n > TTA_{Agent}$ when $T_{(n-1)} < TTA_{Agent}$. Time $T_{(n-1)}$ is when the Requester and the Agent meet for the last time before the Agent collects ads from the Ad-Dealer; therefore, it is smaller than TTA_{Agent}, which is the time that the collection is expected to take place. Time T_n is the time when the Requester and the Agent first meet after the Agent has made the collection; therefore T_n is greater than TTA_{Agent}. Note that if $EA_{(R,A)}$ is greater than TTA_{Agent}, then $T_{(n-1)} = T_0$ and delivery is expected to occur on the following meeting ($DT_{Agent} = T_1$).

In the scenario in Figure 3, an encounter between the Requester and the Agent takes place at $T_0 = 0$. Based on the Encounter Average $EA_{(R,A)}$ which is 10 h, future encounters can be predicted to happen between intervals of approximately 10 h ($T_1 = 10, T_2 = 20, T_3 = 30$). Delivery is expected to occur on the meeting that follows the Agent's visit to the Ad-Dealer that is anticipated to take place 24 h later, as per TTA_{Agent}. Consequently, DT_{Agent} is equal to the time of the first meeting that will take place after a period that is greater than TTA_{Agent}. The $TTA_{Agent} = 24$, therefore, $DT_{Agent} = T_3 = 30$ ($T_3 > TTA_{Agent}$) as the previous encounter $T_2 = 20$ is less than TTA_{Agent}.

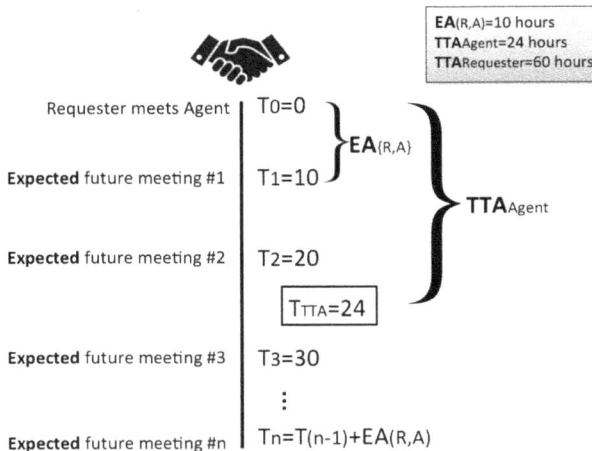

Figure 3. Calculation of DT_{Agent} (Agent's Delivery Time).

The Requester is now able to decide whether or not to use the services of that particular Agent by comparing the Agent's DT_{Agent} with the average time that it takes him/her to collect adverts, i.e., his/her own $TTA_{Requester}$. In this particular scenario, the Requester visits the Ad-Dealer between average time intervals $TTA_{Requester} = 60$ h, which is significantly more than the Agent's $DT_{Agent} = 30$.

Based on this observation, the Requester can deduce that it is within his/her best interest to send his/her requests through that Agent.

5.3.2. UIP Comparison of Requester and Agent

If the Requester decides to use the services of the Agent, he/she then may select a number of specific interests that he/she will request adverts for. In an effort to minimize bandwidth and memory overhead, the Requester attempts to give priority to adverts that the Agent is also interested in. This way, a single advert can be downloaded and viewed by both users.

Interest commonalities could be easily identified by openly comparing the UIPs of the two users, but this would also mean that the users would need to expose their advertising interests to each other. To maintain privacy, the system implements a probability-based UIP comparison mechanism. This mechanism enables the Requester to select some of his/her interests which have a high probability of also being shared with the Agent, but still prevents both of them from learning each other's exact interests.

The UIP, which includes both marked and unmarked interests, is first split into smaller groups of entries, and each of the groups is given a unique name (A, B, C, etc.). The number and size of the groups is determined by the system administrator who is the Broker, and it can vary depending on the total size of the supported UIP, which determines the number of different interests categories that are supported by the system. The size of the groups is a compromise between accuracy and privacy as smaller groups will produce more accurate results while larger groups make it more difficult for a curious Agent to determine the Requester's exact interests. The Agent then composes a list with the group names in descending order based on the number of marked interests in each group and sends it to the Requester. For example, the interest profile UIP that is shown in Figure 4 consists of 50 marked or unmarked interests that are divided into five groups (A, B, C, D and E) with a total of 10 entries in each group. If the Agent were to have the most marked interests in Group C followed by D, then E, A and, finally, B, then his/her composed list L1 would be as L1 = [C,D,E,A,B].

Upon receiving L1, the Requester follows the same procedure with his/her own $UIP_{Requester}$ in order to produce L2 = [B,A,C,D,E]. The Requester can now select a group by simultaneously going through the items in both lists from the top ranked to the bottom ranked until he/she finds a group in L2 that has the same or higher ranking in L1. More specifically, the Requester will begin by comparing the top ranked groups in each list (Group C in L1 and Group B in L2). If they do not match, he/she will then compare the top two items of each list and so forth.

In our particular example, the first comparison will be between L1'= [C] and L2' = [B]. As a match is not found, the second comparison will take place between L1' = [C,D] and L2' = [B,A] that also does not produce a match. The third comparison will be between L1' = [C,D,E] and L2' = [B,A,C] to which the system will identify the leading group to be C, as it has the third highest ranking in L2 and the top ranking in L1. Note that if we had L1' = [A,B] and L2' = [B,A], then the leading group would have been A, as we always select the common item that has the highest ranking in the Agent's list L1.

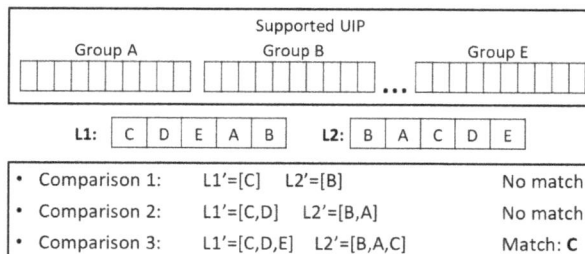

Figure 4. Example of algorithm for User's Interest Profile (UIP) comparison.

Once the leading group has been identified, the Requester picks N marked interests from that group. N is sent to the Requester by the Agent and is the number of requests that the Agent is willing to accept. Should N be greater than the total number of marked interests that the Requester has in the leading group (Group C), then the Requester would continue the algorithm until he/she finds a second group and supplements the remaining interests from there.

5.3.3. Transmission of ARMs

After the Requester has determined which interests to request adverts for, he/she retrieves the corresponding ARMs. Recall that ARMs and the matching keys K_i^{user} have already been composed and stored in memory. The ARMs are labelled with a unique identifier that is called the ARM-ID, and they are ultimately sent to the Agent. Finally, the Requester places his/her copy of the ARMs and K_i^{user} alongside the ARM-IDs, the User-ID of the Agent and a timestamp on a separate queue of pending requests and waits for the requester to return with the requested adverts.

5.4. User Meets Ad-Dealer

Ad-Dealers can dispatch adverts to users within their proximity over anonymous connections through the use of specialized networking equipment, which is similar to that being used in [10,14], which we assume is secure.

5.4.1. Ad-Dealer Authentication and Logging

Before logging their encounter with an Ad-Dealer, users must authenticate the Ad-Dealer. Installed within the user's client is an additional asymmetric verification (public) key K_{auth}, which is the same for all clients, and it is used just for authenticating Ad-Dealers. The matching signing (private) key K_{sign} is given by the Broker to the Ad-Dealers. As shown in Figure 5, users authenticate an Ad-Dealer by sending a random nonce R, which the Ad-dealer must sign with K_{sign}. Since K_{sign} is kept secret among the Ad-Dealers, authentication can be achieved by attempting to verify the signature on the signed R with the use of the K_{auth}.

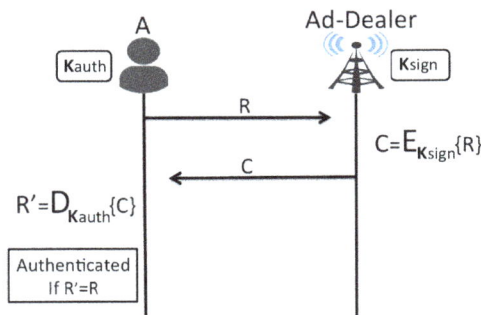

Figure 5. AD-Dealer authentication protocol.

5.4.2. Construction of Bundles and ADMs

After authentication has been successful, the users may log their encounter with the Ad-Dealer and proceed to forward to him/her both the ARMs they constructed themselves, as well as those they received from their Requesters. The AD-Dealer then decrypts the ARMs with DK_{System} and fetches adverts for the requested interests. Depending on availability, each particular interests may be matched to one or multiple ads.

Since the Requester's ARMs are selected after a profile comparison with the Agent, it is to be expected that both users may have multiple requests for the same ads. A significant amount of memory and bandwidth can therefore be conserved by distributing these adverts as a single copy that is addressed to both users. However, this would expose each other's interests and essentially defeat the purpose of performing a private profile comparison in the first place.

Our system bypasses this problem by deploying a novel protocol that enables multiple users to share access to the same ads while preventing them from knowing which ads are being shared. Specifically, the Agent and the Requester are able to view the same encrypted ads, but neither of them can learn what ads are being viewed by the other.

The Ad-Dealer initially encrypts each of the requested adverts with a different asymmetric encryption key EK_i^{AD} (Encryption Key of advert i) of his/her own choosing. The encrypted ads are then stamped with a unique AD-ID number, and they are organized into a sequence of groups that are called Bundles.

The Bundles are composed in such a way that shared ads, i.e., ads that are addressed to multiple users, are mixed with ads that are addressed to the respective users individually. In our particular case, where we only have two users, the Agent and the Requester, a Bundle, which contains a shared ad $AD_{(Agent,Requester)}$ ($AD_{(A,R)}$ for short), also contains some ads that are addressed only to the Agent ($AD_{(A)}^1, AD_{(A)}^2$), as well as some ads that are addressed only to the Requester ($AD_{(R)}^1, AD_{(R)}^2$). Under no circumstances should we have a Bundle that contains only shared ads or a mix of shared ads and ads that are addressed to one of the users, but no ads that are addressed to the other user. The reasons for this become evident later on when we evaluate the privacy of the system.

Once the Bundles have been composed, the Ad-Dealer creates Advert Delivery Messages (ADMs) for every one of the ARMs that he/she received. The ADMs are intended as responses to the ARMs and contain three pieces of information: (1) the AD-IDs of the adverts that relate to interest of a particular ARM, (2) the cryptographic keys DK_i^{AD} (Decryption Key of advert i) that can be used to decrypt them (the ads have been encrypted with the corresponding EK_i^{AD}) and (3) the sequence number of the Bundles in which the adverts are enclosed (note that multiple ads for a single interest may not necessarily be sent in the same Bundle).

Recall that every ARM is labelled with a unique ARM-ID, and it contains a user-generated encryption key K_i^{user}. Each respective ADM is therefore encrypted with the corresponding K_i^{user}, and the resulting ciphertext is labelled with the matching ARM-ID before being sent to the Agent along with all the Bundles of encrypted adverts.

$$ADM_{ARM-ID} = E_{K_i^{user}}\left[AD - ID, DK_i^{AD}, BUNDLE_{(n)}\right] \qquad (3)$$

5.5. User Obtains Ads

5.5.1. Advert Collection from Ad-Dealer

Upon obtaining the ADMs from the Ad-Dealer, the user can use K_i^{Agent} to decrypt the ones that are addressed to him/her, i.e., the ADMs that have the same labels as his/her ARMs. The user can now use the information that was contained in each ADM to locate his/her adverts within the appropriate Bundles and then decrypt them with the DK_i^{AD} that were also contained in the ADM. If the user operates as an Agent, he/she stores the remaining encrypted ADMs that are not addressed to him until he/she gets the chance to deliver them. These are the ADMs that have the same labels as ARMs that he/she forwarded on behalf of a Requester.

5.5.2. Advert Reception through an Agent

When the Agent encounters the Requester after a collection has been made and after the two users have authenticated each other and logged the meeting, the Requester sends an inquiry about the state of the order to which the Agent can respond by transferring the remaining encrypted ADMs. The Requester decrypts the ADMs and obtains the AD-IDs of his/her requested adverts, the sequence numbers of the Bundles they are located in and the DK_i^{AD} keys that can be used to decrypt them.

The Requester can now ask for the specific Bundles, locate his/her ads and then decrypt them. Once the process has been deemed successful, the Requester stores the adverts for future viewing and removes the corresponding ARMs and K_i^{user} from his/her queue of pending requests, where they were placed when the request was made. In the event that an Agent fails to deliver after a certain period of time has passed that is determined by the Requester, then the Requester may reinstate the ARMs from his/her queue of pending requests and proceed to forward them to a different Agent.

6. Evaluation

We evaluate our system by performing a direct security comparison to the relevant work. In Section 3, we examine the currently available systems and identify security risks and shortcomings. The identified shortcomings are taken into account in Section 4.3 where we define a set of operational requirements that will serve as evaluation criteria for our system. For our evaluation, we first determine certain threat scenarios under which the current models would fail to meet our requirements and then use the same scenarios to examine the performance of our own system.

6.1. User Privacy against Other Parties (Broker and Ad-Dealers)

Similarly to the architectures that are described in [11,18], which we assume as secure, our model enables users to establish anonymous connections with a broadcasting station and download content without exposing any identifying information. From that perspective, our system offers a very similar level of protection, as the Ad-Dealer and Broker have no means of obtaining any information that may reveal the user's physical identity, such as username, email and home IP address.

The main difference of our system is that users are also downloading content on behalf of others within their social network. The content is encrypted, and therefore, our system requires the transmission of cryptographic keys, which provides to the Ad-Dealer additional information about the users. The use of cryptography may therefore have implications on privacy, which we will now proceed to examine.

Recall that every ARM that is sent to the Ad-Dealer contains a user-generated key K_i^{user} that is later used by the Ad-Dealer to encrypt the corresponding ADM. Furthermore, the keys are produced by a hash chain where $K_i^{user} = h(K_{i-1}^{user})$. We assume that the Ad-Dealer has no way of obtaining the original key K_0^{user}, which was computed by the user, but knows which hash function $h()$ is used.

Based on this knowledge, the Ad-Dealer can obtain K_i^{user} out of one ARM and then replicate the hash chain in order to compute all consecutive keys K_{i+1}^{user}, K_{i+2}^{user}, etc. without knowing K_0^{user}. This means that the Ad-Dealer can identify ARMs that contain subsequent keys, and from that, he/she may deduce that they have all been created by the same user even if they are sent through multiple Agents and at different time instants. Therefore, the key can be used as an identifier to profile the users. However, this does not compromise privacy to a significant extent as the Ad-Dealer still has no means of learning the user's physical identity. This threat can be minimized if the user changes K_0^{user} very frequently, but this will require the consumption of additional processing power.

One more aspect that needs to be considered is the fact that the Ad-Dealer is also able to infer information about the social interactions of users and, more specifically, the affiliations between Requesters and Agents. For as long as the identities of the users are kept private, this also has minimal implication to privacy, but we do need to consider the possibility of this not always being the case. Should the Ad-Dealer find a way to uncover the true identities of users, then not only can he/she associate it with their advertising interests, but he/she will also learn about their social connections with each other.

The only way of achieving this is by compromising the software client that runs on the user's mobile device. Broker interference with the client would be classified as malicious behaviour; hence, it is out of scope in our case, as we assume an honest-but-curious Broker. Even if the Broker were malicious, tampering with the software would be easily detected and would have serious legal implications. The integrity of the software can be certified by an independent authority such as the developer of the platform it operates on (iOS, Android, Windows mobile).

6.2. User Privacy against Other Users

As adverts are conveyed between peers within a social network, privacy among users is the prime concern that has been mainly dismissed in previously proposed systems. Note that an interaction between two users can be divided into four phases that will now be scrutinized separately.

6.2.1. Requester and Agent Compare UIPs

During the initial phase, a potential Requester and an Agent perform a comparison of their respective interest profiles (UIPs). Recall that the UIP is first separated into groups of entries, and the Agent discloses the group names in descending order based on the number of marked interests he/she has in each group. At this point, the Requester can infer that the Agent has more interests in a high-ranking group than a low-ranked one, but he/she is still completely oblivious to what these interests are or even how many they are. The Agent can assume that the Requester's selection will most likely be from a high-ranking group. Other than this trivial knowledge, the users do not learn any more information about each other's interests, and we can therefore conclude that privacy is preserved.

6.2.2. Agent Receives ARMs from Requester

As has already been stated, ARMs are encrypted with the EK_{System} (System Encryption Key) and, therefore, cannot be read by the Agent, who has no knowledge of the corresponding decryption key DK_{System}. Recall that ARMs contain an Interest-ID, which is unique for the entire system and known to all users, but also contain a key K_i^{user}, which is generated by the user. This essentially means that any two ARMs that contain the same Interest-ID are highly unlikely to produce the same cryptograms as the K_i^{user} is also included as part of the plaintext. An encrypted ARM is therefore secure as it would be infeasible for an attacker to guess its content by comparing it to other encrypted ARMs that he/she creates himself/herself (chosen plaintext attack).

6.2.3. Agent Collects Ads

When appearing within the proximity of an Ad-Dealer, the Agent forwards all of his/her stored ARMs, including his/her own. The Ad-Dealer fetches the requested adverts, encrypts them with different keys of his/her own choosing and he/she organizes them in Bundles. For every ARM, the Ad-Dealer then creates an ADM (Advert Delivery Message), which contains information that will allow the recipient to locate his/her ads within the appropriate Bundle and decrypt them. Finally, the ADMs are encrypted with K_i^{user} and sent back to the Agent. The Agent is able to decrypt the ADMs that are responses to his/her own ARMs, since he/she knows K_i^{Agent}, but the Requester's ADMs remain private as the Agent cannot decrypt them without knowing the $K_i^{Requester}$.

6.2.4. Agent Delivers Ads

Upon meeting the Agent, the Requester first receives his/her ADMs. Then, he/she asks for the Bundles where his/her ads are stored based on the information he/she obtains from the ADMs. Finally, he/she recovers his/her ads, which can either be ads that are meant just for him/her or shared ads, i.e., ads that are addressed to multiple users.

Recall that shared ads are placed within Bundles in such a way that they are mixed with other ads which are addressed to the respective users individually.

Ads that are addressed to only one user, either the Agent of the Requester, remain completely private from the other user who cannot decrypt them without having the appropriate decryption key DK_i^{AD}, which can only be obtained from an ADM. However, the shared ads can be read by both users, and therefore, if one of the users is able to identify which of the ads within a Bundle are shared, he/she will immediately learn the interests of the other user.

Upon receiving a Bundle, the Requester has no means of identifying the shared ads, as they are mixed with the other ads, and therefore, the privacy of the Agent and the privacy of other Requesters who use the same Agent are preserved.

From the Agent's perspective, when the Requester asks for a specific Bundle, then the Agent can assume there is a possibility that some of the ads within the Bundle may be shared. Despite this, the Agent has no way of being absolutely certain of that possibility, as the Requester may be asking for the Bundle just because he/she wants to obtain his/her own ads, which are individually addressed ads that are not visible to the Agent.

If the demanded Bundle were to include no individually-addressed ads for the Requester, the entire content of the Bundle would be visible to the Agent. Then, the Agent would have been able to infer that at least one of the contained ads is shared and that would put the Requester's privacy at risk. For this reason, the Ad-Dealer composes Bundles is such a way that this is never the case. Even in the extreme scenario where the exact same ARMs were sent by both users and, therefore, all of the ads can be shared, the Ad-Dealer prevents this from happening by supplementing additional ads.

One last possibility we need to consider is that a user acts completely maliciously by fabricating his/her UIP, i.e., he/she carefully composes his/her UIP in such a way that only specific interests are marked. A malicious Requester would pose no threat for an Agent as the former does not obtain any information that would enable him/her to identify the shared adverts within a Bundle even if the Requester's UIP is fabricated. A malicious Agent, however, may threaten the privacy of a Requester by employing a fabricated UIP.

Consider a scenario where a malicious Agent engages in a profile comparison with a potential Requester by listing the groups of his/her UIP in order L = [A,B,C,D,E], and the Requester sends him/her two ARMs. At this point, the Agent may assume that it is highly likely that the ARMs are referring to interests that can be found in one of the high-ranked groups, such as A or B. The Agent then composes $ARM_{(x)}$ for a specific interest x, which can be found in one of his/her highly-ranked groups, such as Group A. Afterwards, he/she creates multiple additional ARMs for more interests that belong to low-ranking groups (e.g., Group E or D) and then sends all of them to the Ad-Dealer. If $ARM_{(x)}$ is the same as one of the two ARMs that were sent by the Requester, the Ad-Dealer will retrieve a single shared advert and address it to both users. The Ad-Dealer will also fetch additional adverts for the Requester's second ARM, as well as for the remaining ARMs of the Agent. Since both the Requester and the Agent have at least one individually-addressed advert, the Ad-Dealer can compose all of the ads into a single Bundle and send it back to the Agent. When receiving the Bundle, the Agent should be able to access all of the contained ads except for one: the advert that was requested by the Requester, but not by himself/herself. Based on the fact that the Requester has sent two ARMs, the Agent can infer that within the Bundle, there is one shared advert.

Recall that during the profile comparison operation, the Requester was manipulated into selecting interests out of high-ranking groups (Group A or B), but the Agent requested only advert x out of these groups and all his/her remaining ARM were deliberately selected from low-ranking groups. Based on this, the Agent can deduce that the shared advert concerns interest X.

This type of attack is highly impractical as it is only limited to one interest at a time and cannot work when the Requester sends a larger number of ARMs and the Ad-Dealer responds to each of them with multiple ads that are composed into different Bundles. We can therefore conclude that this attack has a very small chance of being successful, and even when it succeeds, the effects on the Requester's privacy are not very severe.

6.3. Protection from Malicious Content

The spread of fake or harmful content is a serious threat that is not addressed by previously-proposed advert distribution systems that make use of opportunistic technologies. As data are propagated through temporary connections among peers, a malicious Agent could easily replace an advert with his/her own content before forwarding it. Our system overcomes this threat through the use of cryptography on both the ARMs, as well as the ads themselves. As previously stated in Section 5, ARMs are encrypted with EK_{System}, ads are encrypted with EK_i^{AD} and the corresponding DK_i^{AD} is composed into an ADM, which is encrypted with K_i^{user}.

In order for an attacker to replace an advert within a Bundle, he/she would need to encrypt the new malicious advert with the same EK_Z^{AD}, which he/she does not have. If the attacker were to simply use a new encryption key $EK_{Attacker}$ that he/she generates himself/herself, the other users would not be able to decrypt the malicious advert since they do not have the corresponding decryption key $DK_{Attacker}$. For this attack to work, the attacker would first need to create a new ADM that contains the appropriate $DK_{Attacker}$, and then, the new ADM would need to be encrypted with K_i^{user}. Obtaining K_i^{user} however is impossible, as it is enclosed in the ARM that has been encrypted with EK_{System} and can therefore only be accessed by the Ad-Dealer who knows DK_{System}.

One last possibility that needs to be considered is that of the user being tricked into creating ARMs that are not encrypted with EK_{System}. Since EK_{System} is installed within the client software, this could only be possible if the attacker managed to compromise the user's mobile device. If the attacker were to have access to the user's client, then this attack would be trivial since the attacker could easily bypass the system by simply replacing the locally stored adverts. We can therefore conclude that for as long as DK_{System} and EK_i^{AD} are kept secret, it is infeasible for anyone other than an Ad-Dealer to send encrypted adverts to users.

6.4. Protection from Impersonation Attacks

A malicious user can launch an attack against the system by impersonating the identity of a legitimate system stakeholder, either a user or an Ad-Dealer. The four possible scenarios that may accrue from such an attack are the following.

- Victim is a user and the Attacker impersonates one of his/her Contacts (that is not an Agent or a Requester).
- Victim is a Requester and the Attacker impersonates his/her Agent
- Victim is an Agent and the Attacker impersonates his/her Requester
- Victim is an Agent and the Attacker impersonates the Ad-Dealer

In the first scenario, the attacker tricks a user U1 into registering a fake encounter with one of his/her Contacts U2. This would alter the Encounter Average $EA_{(U1,U2)}$ of the two users and in turn interfere with the correct operation of the opportunistic routing algorithm.

In the scenario where the attacker impersonates an Agent there are two cases. In the first case, the victim is a potential Requester, i.e., a user who has not yet made a request, who is manipulated into sending ARMs to the attacker. The attacker would still be unable to read the encrypted ARMs, but this would also result in the requests being lost. In the second case, the victim is already a Requester, i.e., the victim has placed a request and is waiting for ads. If this happens, the Requester may be tricked into downloading data, such as fake ADMs and Bundles, that have not been sent by a legitimate Ad-Dealer.

In the third scenario, the attacker impersonates the Requester, after the request has been made, to an Agent who is tricked into sending to the attacker the adverts (ADMs and Bundles) that are meant for the real Requester. After completing the delivery, the Agent (victim) will have no reason to keep the ADMs, which will be discarded. This will result in a denial of service for the real Requester, who will not be able to obtain his/her adverts when the actual meeting between him and the Agent takes place.

These first three scenarios are prevented with the use of password-based mutual authentication. The authentication protocol uses a standard challenge-response handshake that is assumed to be secure as long as the passwords are not leaked. The original exchange of passwords happens through an out-of-band channel (e.g., manual input, QR code, text) which is very difficult for an attacker to eavesdrop on. Although the passwords are generated automatically by a random function within the client, we assume that they cannot be guessed by an attacker, as the function may include several sources of randomness from the user's device, such as memory usage, time or signal strength.

For our final scenario, the attacker impersonates the identity of an Ad-Dealer. Nearby Agents would be deceived into sending to the attacker their ARMs. The ARMs are encrypted and the attacker would not be able to use them, nor would he/she be able to send back any malicious content. However, the ARMs could potentially be lost, which would result in failure to deliver on behalf of the Agent. Agents avoid this attack by authenticating the Ad-Dealer with the use of digital signatures. The Agent sends a random nonce that the Ad-Dealer is asked to encrypt with the K_{sign} and can be verified with the K_{auth}. We need to note that this creates a threat for the Ad-Dealer being tricked into signing something else other than a user's nonce. However, this would serve no purpose since the Ad-Dealer's signature is used only for this operation.

6.5. Resource Conservation

Wasting of resources is a typical problem of opportunistic networking schemes as multiple copies of the same data exist across the network. Consider a simple example where an Agent receives three different encrypted requests from Requesters R1, R2 and R3 and all of them are for the same advert. When visiting the Ad-Dealer, the Agent would need to collect three copies of the same advert, each copy encrypted with a different key for each Requester. This increases overhead on the Agent side who needs to download and store excess data.

Our system reduces this overhead by taking advantage of the multicast nature of advert delivery as the same advert is relevant for multiple consumers. Instead of circulating multiple copies of the same advert to different users, we are able to spread a single copy that is addressed to multiple recipients. What is innovative about our design is that we achieve this without compromising privacy, as we have already demonstrated in previous sections.

Interest commonalities are expected to emerge naturally among social groups who share the same demographics (age, religion, nationality, etc.) [19]. To further enhance this effect, our system attempts to identify specific shared interests by comparing the user's UIPs. Although it may be counter-intuitive, our aim here is not to achieve complete accuracy. A completely accurate discovery of common interests could compromise privacy among Agents and Requesters. The probabilistic approach that we follow preserves privacy, but at the same time, increases the chances of a randomly selected interest being shared by two users.

This is better illustrated in a simple example. Consider a scenario where a potential Requester wishes to place a request to an Agent. The Agent's interest profile UIP consists of 30 entries out of a total of 100 interests that are supported by the system. This means that if the requester were to perform a random selection, he/she would have a 30% chance of selecting one of the Agent's interests. When we apply our algorithm, the profile is segmented into five groups of twenty interests. The groups are then ranked into descending order which the highest ranked containing 15 interests, the second highest 10, the third five and the two remaining groups containing zero. This enables the Requester to effectively focus his/her priorities on the higher ranked groups and ignore the rest. This would give him a 75% chance of success for the top ranked group, 50% for the second highest ranked group and 25% for the third.

Naturally, these numbers will vary depending on how sparsely the Agent's interests are spread across the UIP. The total number and size of the groups can be altered depending on the number of supported interests and the desired level of accuracy. In addition to this, the groups can potentially be composed based on logical criteria, e.g., each group containing interests that are relevant to specific types of consumers. For example, a group may include interests that are commonly associated with male consumers while another group will include interests that are common among females. This is very similar to the profiling method that is currently used by advertising companies and will result in greater accumulation of marked entries within specific groups. However, in order to maintain privacy, the groups need to be created based on criteria that do not expose any information that the Requester and Agent do not already know about each other. This may include demographic information that is very generic such as gender, age and nationality.

7. Conclusions and Future Work

In this paper, we present an advert distribution system, which combines ad-broadcasting (anonymous download of ads from broadcasting stations) with opportunistic networks. We first classify and scrutinize relevant systems in terms of privacy, security, targeting effectiveness and practicality and then recognize their shortcomings. For our evaluation, we use the identified shortcomings of relevant systems in order to examine the effectiveness of our own system under different threat scenarios. Our evaluation shows that our system achieves a greater level of privacy by eliminating the need for trust among users, which are nodes of the opportunistic network, and at the same time, reduces memory and bandwidth overhead by allowing multiple users to share access to the same ads without compromising their privacy. Counter to previous models, our system considers the possibility of malicious activity and effectively prevents the spread of fake ads from a rogue node, as well as impersonation attacks.

We are currently in the development stage of a working prototype, which will enable us to experimentally assess the delivery performance and the resource efficiency of our model, including memory footprint and impact on battery. Additionally, we intend to incorporate a secure click-report mechanism that will prevent fraud against the Advertisers, eliminate the need for a Broker and, at the same time, work as a reward system, which will provide incentives for the participation of Agents and Ad-Dealers.

Author Contributions: Stylianos S. Mamais, security design and manuscript preparation. George Theodorakopoulos, initial idea, suggestions on cryptography.

Conflicts of Interest: The authors declare no conflict of interest.

References

1. eMarketer. Mobile to Account for More than Half of Digital Ad Spending in 2015. Available online: https://www.emarketer.com/Article/Mobile-Account-More-than-Half-of-Digital-Ad-Spending-2015/1012930 (accessed on 18 January 2017).
2. The Guardian. UK Mobile Ad Spend 'To Overtake Print and TV'. Available online: https://www.theguardian.com/media/2015/sep/30/mobile-advertising-spend-print-tv-emarketer (accessed on 18 January 2017).
3. Yan, J.; Liu, N.; Wang, G.; Zhang, W.; Jiang, Y.; Chen, Z. How much can behavioural targeting help online advertising? In Proceedings of the 18th International Conference on World Wide Web, Madrid, Spain, 20–24 April 2009; pp. 261–270.
4. Purcell K.; Brenner J.; Rainie L. Pew Research Center: Search Engine Use 2012. Available online: http://www.pewinternet.org/2012/03/09/search-engine-use-2012/ (accessed on 18 January 2017).
5. Federal Trade Commission (FTC). FTC Staff Report: Self-Regulatory Principles for Online Behavioural Advertising: Tracking, Targeting, and Technology. Available online: https://www.ftc.gov/sites/default/files/documents/reports/federal-trade-commission-staff-report-self-regulatory-principles-online-behavioural-advertising/p085400behavadreport.pdf (accessed on 18 January 2017).
6. Sun, Y.; Ji, G. Privacy preserving in personalized mobile marketing. In *Active Media Technology: 6th International Conference, AMT 2010, Toronto, Canada, August 28–30, 2010 Proceedings*; Springer: New York, NY, USA, 2010; pp. 538–545.
7. Backes, M.; Kate, A.; Maffei, M.; Pecina, K. Obliviad: Provably secure and practical online behavioural advertising. In Proceedings of the 2012 IEEE Symposium on Security and Privacy, San Francisco, CA, USA, 21–23 May 2012; pp. 257–271.
8. Chor, B.; Goldreich, O.; Kushilevitz, E.; Sudan, M. Private information retrieval. In Proceedings of the IEEE 36th Annual Symposium on Foundations of Computer Science, Milwaukee, WI, USA, 23–25 October 1995; pp. 41–50.

9. Toubiana, V.; Narayanan, A.; Boneh, D.; Nissenbaum, H.; Barocas, S. Adnostic: Privacy preserving targeted advertising. In Proceedings of the Network and Distributed System Symposium, San Diego, CA, USA, 28 February–3 March 2010.

10. Kodialam, M.; Lakshman, T.; Mukherjee, S. Effective ad targeting with concealed profiles. In Proceedings of the IEEE International Conference on Computer Communications (INFOCOM), Orlando, FL USA, 25–30 March 2012; pp. 2237–2245.

11. Guha, S.; Cheng, B.; Francis, P. Privad: Practical privacy in online advertising. In Proceedings of the 8th USENIX conference on Networked Systems Design and Implementation, Boston, MA, USA, 30 March–1 April 2011; pp. 169–182.

12. Carrara, L.; Orsi, G. *A New Perspective in Pervasive Advertising*; Technical Report; Department of Computer Science, University of Oxford: Oxford, UK, 2011.

13. Carrara, L.; Orsi, G.; Tanca, L. Semantic pervasive advertising. In Proceedings of the 7th International Conference on Web Reasoning and Rule Systems, Mannheim, Germany, 27–29 July 2013; pp. 216–222.

14. Straub, T.; Heinemann, A. An anonymous bonus point system for mobile commerce based on word-of-mouth recommendation. In Proceedings of the 2004 ACM Symposium on Applied Computing, Nicosia, Cyprus, 14–17 March 2004; pp. 766–773.

15. Ratsimor, O.; Finin, T.; Joshi, A.; Yesha, Y. eNcentive: A framework for intelligent marketing in mobile peer-to-peer environments. In *The 5th international conference on Electronic Commerce (ICEC 2003)*; ACM: New York, NY, USA, 2003; pp. 87–94.

16. Ntalkos, L.; Kambourakis, G.; Damopoulos, D. Let's Meet! A participatory-based discovery and rendezvous mobile marketing framework. *Telemat. Inform.* **2015**, *32*, 539–563.

17. Lindgren, A.; Doria, A.; Schelen, O. Probabilistic routing in intermittently connected networks. In *Service Assurance with Partial and Intermittent Resources*; Springer: New York, NY, USA, 2004; pp. 239–254.

18. Haddadi, H.; Hui, P.; Brown, I. MobiAd: Private and scalable mobile advertising. In Proceedings of the Fifth ACM international Workshop on Mobility in the Evolving Internet Architecture, Chicago, IL, USA, 20–24 September 2010; pp. 33–38.

19. Lazer, W. *Handbook of Demographics for Marketing & Advertising: New Trends in The American Marketplace*; Lexington Books: New York, NY, USA, 1994.

future internet

MDPI

Article

User Modelling Validation over the Security Awareness of Digital Natives

Vasileios Gkioulos [1,*,†], Gaute Wangen [1,†] and Sokratis K. Katsikas [1,2,†]

[1]   Department of Information Security and Communication Technology,
       Norwegian University of Science & Technology, 2802 Gjøvik, Norway;
       gaute.wangen@ntnu.no (G.W.); sokratis.katsikas@ntnu.no (S.K.K.)
[2]   Department of Digital Systems, University of Piraeus, 18532 Piraeus, Greece
*   Correspondence: vasileios.gkioulos@ntnu.no; Tel.: +47-6113-5162
†   These authors contributed equally to this work.

Academic Editor: Georgios Kambourakis
Received: 7 June 2017; Accepted: 5 July 2017; Published: 10 July 2017

Abstract: Young generations make extensive use of mobile devices, such as smart-phones, tablets and laptops, for a variety of daily tasks with potentially critical impact, while the number of security breaches via portable devices increases exponentially. A plethora of security risks associated with these devices are induced by design shortcomings and vulnerabilities related to user behavior. Therefore, deploying suitable risk treatments requires the investigation of how security experts perceive the digital natives (young people, born in the digital era), when utilizing their user behavior models in the design and analysis of related systems. In this article, we present the results of a survey performed across a multinational sample of security professionals, in comparison to our earlier study over the security awareness of digital natives. Through this study, we seek to identify divergences between user behavior and the conceptual user-models that security experts utilise in their professional tasks. Our results indicate that the experts understanding over the user behaviour does not follow a solidified user-model, while influences from personal perceptions and randomness are also noticeable.

Keywords: security; mobile devices; digital natives; security awareness; security experts; user behavior; education

1. Introduction

Mobile devices tend to become an indispensable part of our everyday life, by fulfilling the increasing user need for access to services and information, without time or location related restrictions. Therefore, the proliferation of such devices and increased user dependency promoted the transmutation of mobile devices to multifunctional equipment, where their increasing computational and storage capacity allow them to provide critical services with significant security implications. Accordingly, mobile devices (i.e., cell phones, tablets, and laptops) store and process critical information, associated with their owners but also people, legal entities and infrastructure related to them.

Furthermore, users tend to seek access to services being unaware or uninterested on the potential associated risks [1–3], while they become accustomed to continuous connectivity. Therefore, such practices are common even across networks with unknown configurations, while they are likely to leave users vulnerable to unauthorized access, allowing associated risks to materialize into attacks. Therefore, vulnerabilities arising from user-behavior or design shortcomings can facilitate malicious activity, allowing adversaries to launch attacks that can lead to privacy breaches and identity theft. In light of this actuality, it is important that users are aware of the associated risks, and, more importantly, that the developed systems and services are adjusted to realistic user models.

This article builds on our earlier study on the security awareness of digital natives [4], where we examined the behavior of this group with respect to their educational background and levels of security competence. This initial study allowed us to establish an understanding and extract related findings over the security awareness and behavior of digital natives within four focus areas: (i) Use of Mobile Devices; (ii) Connectivity and Network Access; (iii) Management of Credentials; and (iv) Protection mechanisms. Specifically, this study investigates how well the security experts manage to predict the user behavior of the digital natives within the four areas above.

Nevertheless, due to the intrusive nature of these technologies and increasing dependency for the execution of daily tasks, educating the users over security-related best practices can only take us halfway towards securing this environment. One of the findings of the aforementioned study was that digital natives are willing to compromise access to services in favor of security, when they are provided with usable solutions. Therefore, it is crucial to investigate how users are modeled during the design or analysis phases of such systems since deviations between user-behavior and user-models can promote security analyses and design decisions with opposite results. Consequently, this article seeks to utilize the findings of our initial study, in order to investigate the conceptual understanding of security experts with respect to the user behavior of digital natives.

The remaining of the paper is structured as follows: Section 2 presents related work, while Section 3 sets the scope of this study by identifying the utilized methodologies and sample areas. Consequently, Section 4 presented the results across the aforementioned focus areas, and Section 5 provides a summary of results and discussion. The article continues with a reference to the limitations, suggestions for future work and the conclusions.

2. Related Work

Since 2014, mobile devices are well on their way to becoming the leading digital platform, displacing the desktop PC [5]. Prensky [6] writes that the digital natives have radically changed their way of thinking by being exposed to technology almost since birth, while other scholars [7] have contested such claims. However, there is no denying that the digital natives have a different view of technology than older generations. This section summarizes the related work gathered on user behavior and modeling, user model validation, and, lastly presents the specific studies this paper builds on.

A considerable number of studies have been conducted on user behavior with regards to selecting and installing applications in smartphones: one found that users do not consider security and privacy issues during app selection, by ignoring privacy policies and EULAs (End-user license agreement) [8]. In addition, Android users were found not to be knowledgeable about permission information during installation [1,2]. Furthermore, studies of user security awareness are useful for understanding and modeling the Digital Natives' behavior. One such study explored the security awareness metrics of smartphone users and found that the security background had a slight effect on the smartphone security awareness of their sample [9]. A second study of security awareness found that users who download applications from official application repositories are complacent in their smart-phone security behaviors and display high levels of trust towards smart-phone application repositories [10]. In addition, they rarely consider privacy and security when installing new applications, and do not install adequate protection mechanisms [3]. Additional research into users and protection mechanisms partially contradicts that smartphone users are not security aware, and finds several correlations between security awareness and smartphone OS, language, and gender [11]. However, the results do not specifically target the digital natives, despite them being the majority of users. Two studies carried targeting the Slovenian digital natives found that the student population had a low awareness of security threats and security measures [12,13].

These studies show that there exist multiple studies of the digital natives' security awareness and behavior. The contradicting results also suggest that culture and background is a variable that should be considered when researching and modeling the digital natives. Given the value of being able to predict human behavior, there are also multiple studies from information security that attempts to model the adversary [14,15], employee behavior [16], and generic user behavior [17]. Validation studies of mental models show that there is a gap between the mental models of security experts and non-experts [18,19]. However, our literature survey found no research on validating the security expert predictions regarding the behavior of digital natives. This is an important topic as it is the security experts that are designing the mechanisms being implemented into the devices, and if the security mechanisms are misaligned with the user group, it can impede both the security and the user experience [20].

Ariu et al. [21] have worked on filling this gap by studying the level of awareness and perception of IT security among university students, paying particular attention to the world of mobile devices. Their report analyses the answers given by 1012 students from over 15 Italian universities to a multiple-choice questionnaire. This shows that students' perception of their knowledge is wrong and that they are unaware of the risks arising from their behavior. The current paper builds on the Ariu et al. results and supplements with two additional datasets from a study collected by Gkioulos et al. [4]. The latter study presents a second data set collected from generic computer science students and a third dataset collected specifically from information and cyber security students. The Gkioulos et al. study highlights several differences between the three groups attributed to security education and awareness, but also commonalities across them. These results suggest that users tend to demonstrate negligible behaviour in the daily use of their mobile devices, due to increased confidence in their security related competence. Additionally, digital natives remain unaware of the full extent of countermeasures that are available at their disposal, while they prioritize access to services and usability over the enforcement of security measures. Furthermore, digital natives are willing to accept security related risks despite their concerns, while they feel less constrained when they are using laptops in comparison to smart-phones and tablets. The full extent of these initial results and datasets are applied in the current study for the user modelling validation of the digital natives.

3. Methodology

This article builds on a previous study [21] on the topic of security awareness of the digital natives, which was conducted to investigate the differences in risk perception across three distinct groups categorized by their technical background. This section has the following structure: the first sub-section addresses the choice of data collection method and instrument, followed by the sample description, and a brief overview of the statistical methods used for data analysis.

3.1. Data Collection and Instrument

The initial data collection [4] aimed to explore the security awareness of the digital natives addressed to students of the digital age, i.e., persons who were born in the years of the technological boom in Information Technology and Communications (ICT), between 1987 and 1997. The three data samples of the digital natives groups were collected from European universities as these are ideal since they comprise a diverse population. We found the online questionnaire to be the best option for data gathering as it reaches a broad audience and provides a strong level of anonymity; therefore, the presented datasets were collected using Google Forms.

The original survey was developed by Ariu et al. [21] and initially ran in multiple Italian universities [21]. The survey had 60 questions that investigated security awareness aspects within the five areas outlined in the introduction. As for the level of measurement, the questionnaire had category, ordinal, and continuous type questions. Category type questions are used here mainly for demographics, while the main bulk of the questionnaire was designed using several mandatory scales and ranking questions.

We designed a second questionnaire to identify divergences between user behavior and the conceptual user models that security experts utilize in their professional tasks. This survey had 44 questions designed to see how well the expert could predict the responses of the three initial digital natives group. The issues in the second questionnaire had a similar layout as the original. However, each question was framed such that the expert predicted what each original sample group answered to a question.

3.2. Sample Description

This study utilizes three distinct samples of security competence groups (SCGroups) collected previously [4,21]:

- The data set collected by Ariu et al. targeted to the Italian digital natives is included in this study and corresponds to our general security competence group (GSCG). The sample consists of 1012 respondents from various university departments (including law, engineering, computer science, humanitarian, marketing, and multiple other faculties not directed to IT education), which we map to the general population of digital natives for this study.
- Secondly, we collected data for the medium security competence group (MSCG) by targeting digital natives from Greece with education exclusively in computer science [4]. We expected this group to have a wider knowledge of the use of mobile technologies and increased awareness over security related aspects due to their educational background. The sample consists of 303 responds, categorized on undergraduate (234), postgraduate (54), and doctoral (15) levels.
- Finally, the target population for the high-security competence group (HSCG) was undergraduate, postgraduate and doctoral students of information security from Norway [4]. These were expected to have a higher security awareness regarding the four main areas than the medium and generic groups, due to their specialized education. For this group, we had 35 respondents in total, of which 21 are undergraduate students, ten postgraduate, and four doctoral students.

Furthermore, the data collection for the second questionnaire targeted experts within information and cyber security. An expert in this study has worked within the field for at least five years, together with a specific skill or knowledge set relevant to our target group. The invited participants of the study had internationally recognized expertise demonstrated through either: (i) a record of scientific publications on security issues, preferably in peer-reviewed publications; (ii) experience at a high level in global, regional, or national security assessments; and (iii) experience at a high level in the design and management of security. Based on these criteria, we distributed primarily to European experts known to the researchers. In total, we contacted 166, of which we received 34 answers (20.5% response rate), distributed as follows: industry (10), academia (19), and security developers, programmers, or similar (5).

3.3. Analysis

The difference in the number of respondents for each survey reflects the scarcity of each group in the general population, for example, there are more respondents in GSCG than MSCG and more in MSCG than HSCG. Both the HSCG and the expert groups being large enough for the central limit theorem to apply [22]. The questionnaire primarily asked categorical and ordinal multiple-choice questions, while, as a measurement of central tendency for ordinal questions, we considered the median, variance, and range. For the descriptive data analysis, we primarily consider differences in the frequency distributions, while we used the security competence groups as categorical data for bi-variate analysis in the original dataset.

In order to measure how well the experts managed to predict the digital natives behavior, we compared the original datasets with the expert responses. We used the original percentages to indicate concrete yes/no questions and the median to indicate the expert response. For example, we asked the experts to predict how many out of ten they thought answered yes to a specific question derived from the initial results (1st round results) [4] and then we used the median and skewness of the expert prediction to compare. We have used the right answer ± 1 (10%) as a correct expert prediction of the user behavior in the analysis.

Future Internet **2017**, *9*, 32

To summarize the results and obtain a measure of expert prediction accuracy, we analyzed each prediction included in this study, totaling 27 for the GSCG and MSCG groups, and 29 for the HSCG. Furthermore, we judged a successful prediction of the outcome of the rating questions when the experts predicted a median within ±1 of the first round digital natives results [4]. This result corresponds to a total 20% deviation around the central value from the original results and 10% in the cases where the initial result is 0 or 10. For the multiple choice questions, a successful prediction equaled a majority expert vote on the most frequently chosen alternative from the digital natives study. The sum of all correct and missed predictions equals the total accuracy.

4. Analysis of Results and Discussion

This section presents the results of our study, where security specialists estimated the responses of the digital natives from the first stage [4] in accordance with their conceptual user models. The comparative analysis aims to identify divergences between the responses of the digital natives and the expectations of the security professionals.

4.1. Use of Mobile Devices

4.1.1. Question 1

We asked the security experts to identify how many out of ten digital natives across the three SCG groups restore the factory settings of their mobile device prior to selling or donating it. The results presented in Figure 1 show that the security experts underestimated the responses of the GSCG group (representative of the general digital natives population), with a difference of −3.27 and only 29.4% of them approximating it correctly (8 ± 1). Additionally, the distribution of responses in respect to the GSCG group is not concentrated around a central value, with a skewness of 0.0446 and a median deviation equal to 2. However, the results in respect to the MSCG/HSCG groups are improved in comparison, with 44.1% and 58.8% of the experts approximating the results correctly, while the values of difference and skewness are also improved.

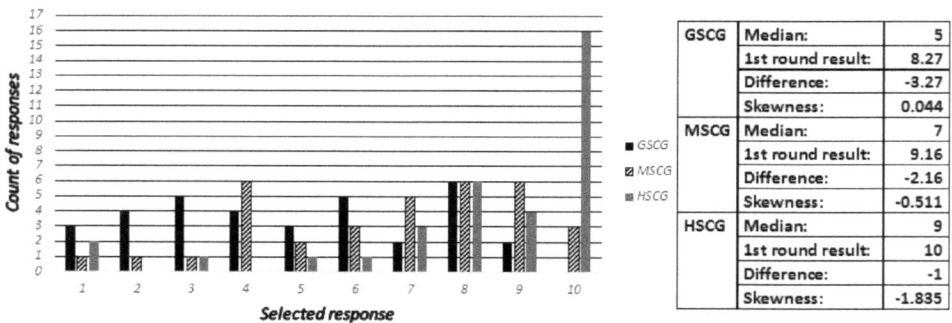

GSCG	Median:	5
	1st round result:	8.27
	Difference:	-3.27
	Skewness:	0.044
MSCG	Median:	7
	1st round result:	9.16
	Difference:	-2.16
	Skewness:	-0.511
HSCG	Median:	9
	1st round result:	10
	Difference:	-1
	Skewness:	-1.835

Figure 1. Results for question 1. Graph on the left shows distribution of expert predictions and the table to the right shows comparison results.

4.1.2. Questions 2 and 3

Focusing on software updates, we asked the digital natives "Do you regularly update the software of your mobile device?", maintaining two distinct categories for cell-phones/tablets and laptops. The possible answers were: "Applications and operating system", "Only applications", "Only operating system" and "No, I do not update". Subsequently, the security experts were asked to identify which was the most frequent and second most frequent answer across the three SCG groups. The responses of the digital natives are presented in Table 1, and the corresponding responses of the security experts in Figures 2 and 3. The results show that the expectations in respect to the responses of the GSCG and MSCG groups are again underestimated, while the lack of a central value is visible, especially for the case of the GSCG group.

Table 1. First round results for questions 2 and 3; GSCG: general security competence group; MSCG: medium security competence group; HSCG: high-security competence group.

Software Updates for Cell-Phones/Tablets		
	Most Frequently	**Second Most Frequently**
GSCG	Apps and OS, 81.3%	Apps, 8.8%
MSCG	Apps and OS, 79.5%	Apps, 8.3%
HSCG	Apps and OS, 88.6%	Apps, 8.6%
Software Updates for Laptops		
GSCG	Apps and OS, 75.3%	I do not update, 14.5%
MSCG	Apps and OS, 73.1%	I do not update, 8.0%
HSCG	Apps and OS, 94.1%	Apps, 5.9%

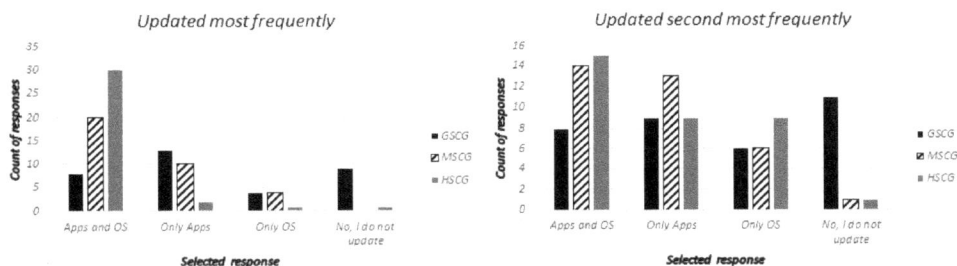

Figure 2. Expert predictions for question 2—cell-phone/tablets.

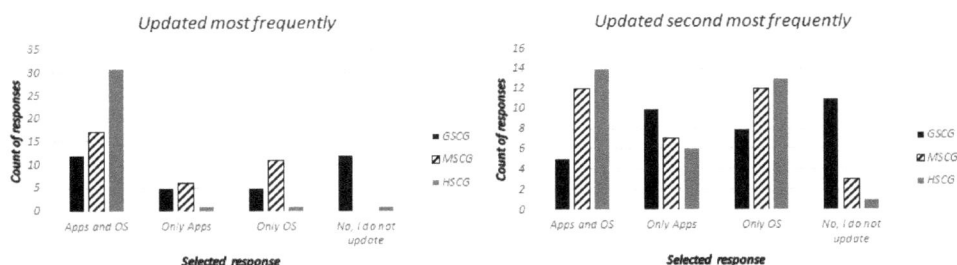

Figure 3. Expert predictions for question 3—laptops.

4.1.3. Question 4

We asked the digital natives "How frequently do you check the permissions (access rights) that an application requires before completing the installation", with possible responses being "Never", "Rarely", "Often" and "Always". Accordingly, the security experts have been asked to estimate the most frequent and second most frequent responses across all SCG groups, as well as how many out of ten digital natives have selected "Always" and "Never". The responses are presented in Figures 4–6, while the most frequent answers of the digital natives from the first round of results have been: GSCG—(Rarely: 38.2%, Often: 25.5%), MSCG—(Always: 34.9%, Often: 25.5%), HSCG—(Always: 40%, Often: 40%). The results show that the experts missed both options for the GSCG, got the second most frequently right for the MSCG, and both options right for the HSCG.

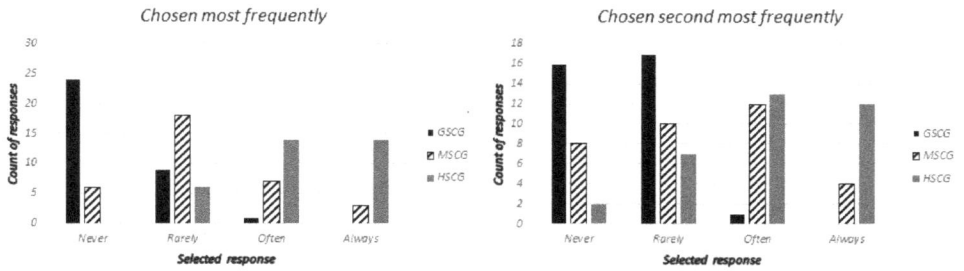

Figure 4. Expert predictions for question 4: "Chosen most" and "Chosen second most" frequently.

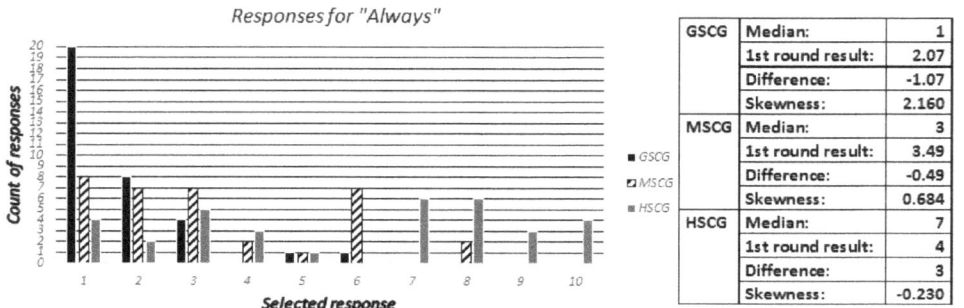

Figure 5. Expert predictions for question 4: Distribution of "Always"-response and comparison results.

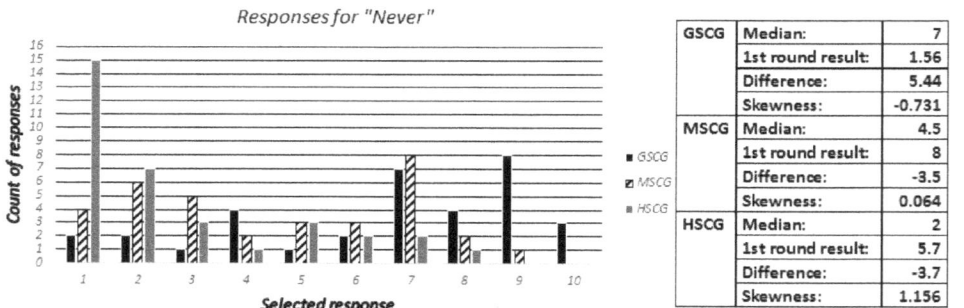

Figure 6. Expert predictions for question 4, Distribution of "Never"-response and comparison results.

4.1.4. Question 5

We asked the security professionals to estimate how many out of ten participants across all groups would report to the authorities the loss or theft of their mobile device. Similar to that of previous questions, Figure 7 shows a noticeable difference where the expert groups significantly underestimate the behavior of the GSCG group.

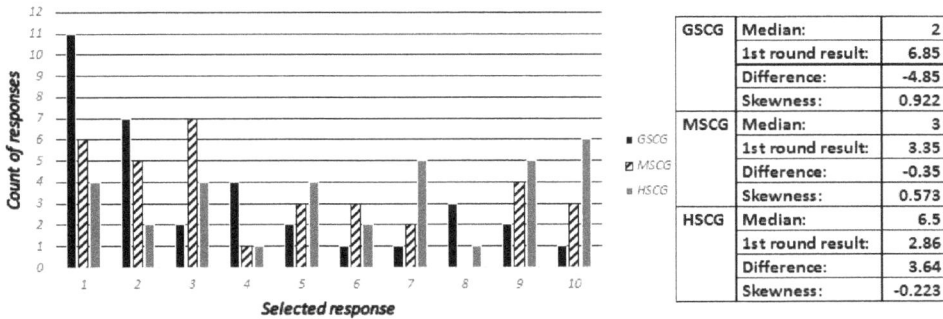

GSCG	Median:	2
	1st round result:	6.85
	Difference:	-4.85
	Skewness:	0.922
MSCG	Median:	3
	1st round result:	3.35
	Difference:	-0.35
	Skewness:	0.573
HSCG	Median:	6.5
	1st round result:	2.86
	Difference:	3.64
	Skewness:	-0.223

Figure 7. Expert predictions for question 5.

4.1.5. Question 6

Aiming to isolate the results corresponding to the digital natives with security related background, we asked the experts to estimate how many out of ten members of the HSCG group pay attention to the signs of a secure connection, when using their laptop for activities that require a high level of security. Question 6a (Q6a) asked experts to predict how many out the HSCG answered "Always" or "Often", and Q6b asked how many answered "Sometimes", "Rarely", or "Never". According to the results presented in Figure 8, the differences between the expectations of the security experts predictions and the responses of the digital natives are minimal for this question.

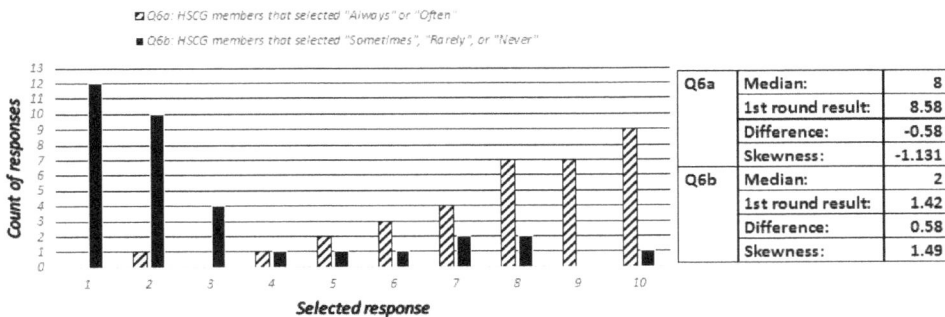

Q6a	Median:	8
	1st round result:	8.58
	Difference:	-0.58
	Skewness:	-1.131
Q6b	Median:	2
	1st round result:	1.42
	Difference:	0.58
	Skewness:	1.49

Figure 8. Expert predictions for question 6, only for HSCG (high-security competence group).

4.2. Connectivity and Network Access

Question 7

We asked the digital natives about their behavior when they discover an open Wi-Fi access point with their cell-phone/tablet or their laptop. The available answers were "I connect and use all applications", "I connect but only do activities that do not require credential authentication", and "I do not connect". Accordingly, the security experts have been asked to estimate the most frequent response, and how many out of ten participants have selected the answer "I connect and use all applications" across the three SCG groups. The security professionals correctly identified the most frequent response in five out of the six scenarios (3 SCG groups * 2 Types of devices), except the response of the HSCG group for cellphones/tablets where "I do not connect" was the most frequent answer by 42.9%, while the estimation promoted the "I connect but only do activities that do not require credential authentication" response. Nevertheless, as presented in Figure 9, the expectations in respect to the security awareness of the digital natives are significantly overestimated especially for the GSCG group.

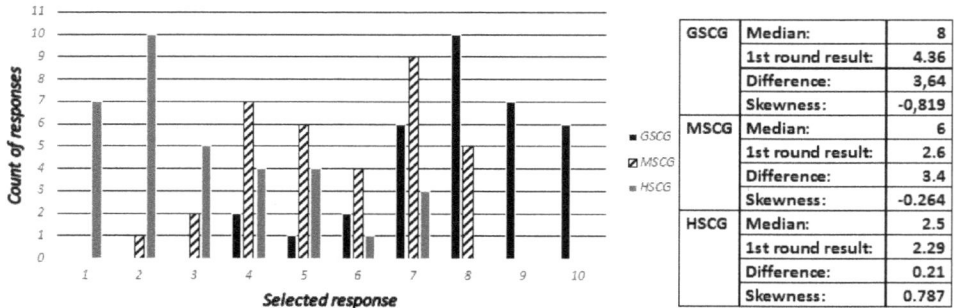

GSCG	Median:	8
	1st round result:	4.36
	Difference:	3,64
	Skewness:	-0,819
MSCG	Median:	6
	1st round result:	2.6
	Difference:	3.4
	Skewness:	-0.264
HSCG	Median:	2.5
	1st round result:	2.29
	Difference:	0.21
	Skewness:	0.787

Figure 9. Expert predictions for question 7.

4.3. Management of Credentials

4.3.1. Question 8

The security professionals have been asked to estimate how many out of ten digital natives across all SCG groups store personal passwords as plaintext in their mobile devices. It is noticeable from the results presented in Figure 10, which shows that an average of approximately 29% of digital natives follow this practice regardless of their security competence or background, while the security experts overestimated the results in respect to the GSCG/MSCG groups.

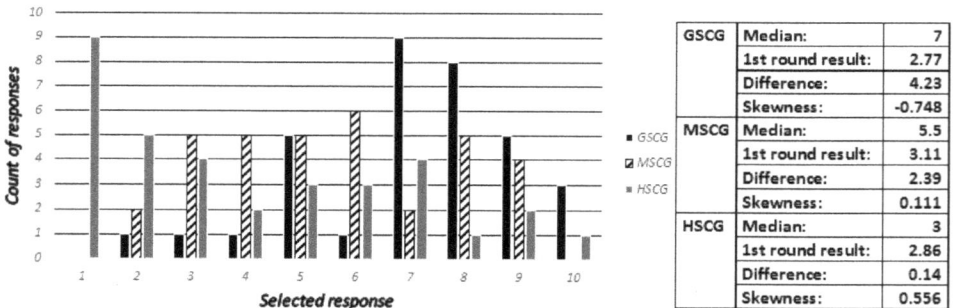

GSCG	Median:	7
	1st round result:	2.77
	Difference:	4.23
	Skewness:	-0.748
MSCG	Median:	5.5
	1st round result:	3.11
	Difference:	2.39
	Skewness:	0.111
HSCG	Median:	3
	1st round result:	2.86
	Difference:	0.14
	Skewness:	0.556

Figure 10. Expert predictions for question 8.

4.3.2. Question 9

Consequently, we asked the digital natives "As soon as you have finished using an application, you...?", with available answers being "Save credentials to stay logged in", "Log out", "Forget to log out", "Do not log out because it is not important" and "Do not log out because I do not know how". Accordingly, the security experts have been asked to estimate how many out of ten participants across the three groups have selected the most critical regarding security out of these responses. Figure 11 shows the expert predictions of option "Save credentials to stay logged in", which shows that the experts overestimate the frequency of saving the credentials for the GSCG and the MSCG while underestimating the HSCG. Figure 12 indicates that all three competence group have very similar behaviors for logging out of applications, while the experts falsely predicted a higher frequency of logging with increased security competence. Figure 13 shows the results for "Do not log out because it is not important" where all three competence groups have answered within the range of 1.08–2.05, whereas the experts overestimated the GSCG and the MSCG, and correctly predicted the HSCG.

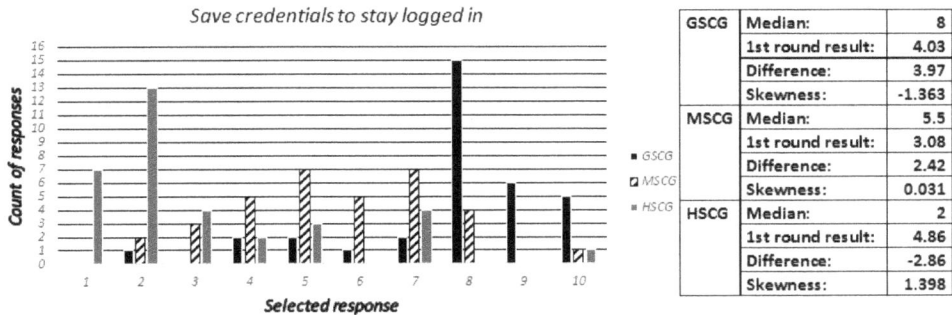

GSCG	Median:	8
	1st round result:	4.03
	Difference:	3.97
	Skewness:	-1.363
MSCG	Median:	5.5
	1st round result:	3.08
	Difference:	2.42
	Skewness:	0.031
HSCG	Median:	2
	1st round result:	4.86
	Difference:	-2.86
	Skewness:	1.398

Figure 11. Expert predictions for question 9, "Save credentials to stay logged in".

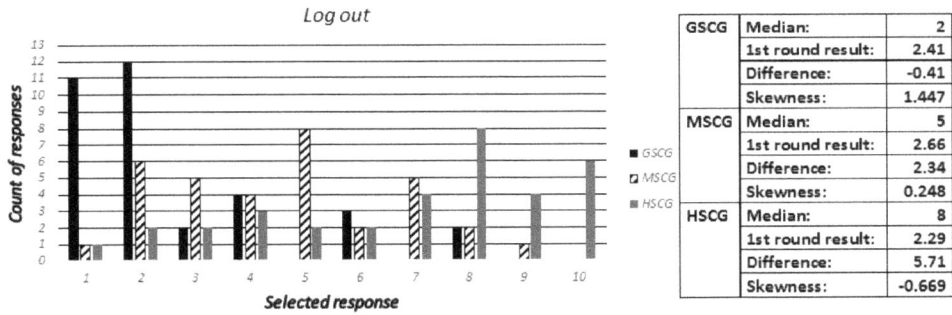

GSCG	Median:	2
	1st round result:	2.41
	Difference:	-0.41
	Skewness:	1.447
MSCG	Median:	5
	1st round result:	2.66
	Difference:	2.34
	Skewness:	0.248
HSCG	Median:	8
	1st round result:	2.29
	Difference:	5.71
	Skewness:	-0.669

Figure 12. Expert predictions for question 9, "Log out".

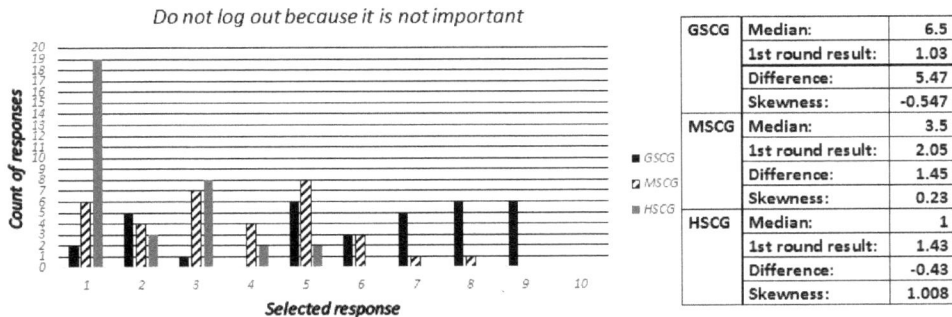

GSCG	Median:	6.5
	1st round result:	1.03
	Difference:	5.47
	Skewness:	-0.547
MSCG	Median:	3.5
	1st round result:	2.05
	Difference:	1.45
	Skewness:	0.23
HSCG	Median:	1
	1st round result:	1.43
	Difference:	-0.43
	Skewness:	1.008

Figure 13. Expert predictions for question 9, "Do not log out because it is not important".

4.4. Protection Mechanisms

4.4.1. Question 10

The security professionals have been asked to estimate the two most favorable access control methods across the three digital natives groups. These included "Biometrics", "Pass-phrases", "Pattern lock", "PIN" and "None". Furthermore, they have been asked to estimate how many out of ten digital natives stated that they do not utilize any access control method for their mobile devices. The results presented in Table 2 and Figure 14 show that the security experts have a sufficient understanding of the technology penetration rates for the available access control methods across the population and the extent of their utilization.

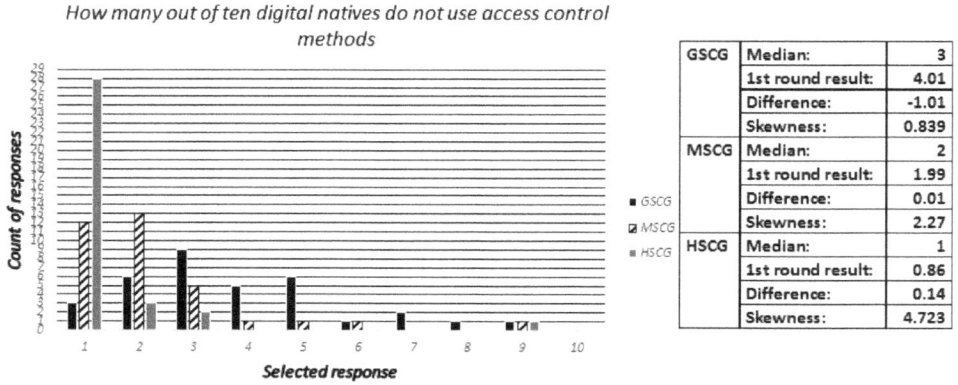

<table>
<tr><td>GSCG</td><td>Median:</td><td>3</td></tr>
<tr><td></td><td>1st round result:</td><td>4.01</td></tr>
<tr><td></td><td>Difference:</td><td>-1.01</td></tr>
<tr><td></td><td>Skewness:</td><td>0.839</td></tr>
<tr><td>MSCG</td><td>Median:</td><td>2</td></tr>
<tr><td></td><td>1st round result:</td><td>1.99</td></tr>
<tr><td></td><td>Difference:</td><td>0.01</td></tr>
<tr><td></td><td>Skewness:</td><td>2.27</td></tr>
<tr><td>HSCG</td><td>Median:</td><td>1</td></tr>
<tr><td></td><td>1st round result:</td><td>0.86</td></tr>
<tr><td></td><td>Difference:</td><td>0.14</td></tr>
<tr><td></td><td>Skewness:</td><td>4.723</td></tr>
</table>

Figure 14. Expert predictions for question 10.

Table 2. Expert predictions for question 10.

Digital Natives Responses		
	Most Frequent	**Second Most Frequent**
GSCG	None, 40.1%	PIN, 28.8%
MSCG	Pattern lock, 34.9%	PIN, 31.4%
HSCG	Biometrics, 54.3%	Pass-phrase, 48.6%
Security Experts Estimation		
GSCG	Pattern lock, 38.2%	PIN, 32.3%
MSCG	Pattern lock, 35.3%	Biometrics/Pass-phrase, 23.5%
HSCG	Biometrics, 52.9%	Pass-phrase, 20.5%

4.4.2. Question 11

Similarly to question 10, we asked the security experts to identify the most favorable protection tools used by the digital natives across all groups. The available responses included "Lock wipe", "Remote wipe", "Find my phone", "Backup", "Encryption", "Personal firewall", "VPN", "None", "I do not know these tools" and "Other". The most favorable and second most favorable tools selected among the digital natives across the three groups are presented in Table 3, along with the estimations of the security experts. The results show that the experts got one out of six possible right (the second most frequent for GSCG). Furthermore, the results with respect to the critical "None" and "I do not know these tools" responses are presented in Figures 15 and 16. The results show that for the "None" response the experts estimated a too high median for the GSCG and correctly predicted the two other groups. Secondly, all the digital natives groups had good knowledge of the tools, but the experts underestimated the GSCG and MSCG groups knowledge by predicting too high values.

Table 3. Expert predictions for question 11.

Digital Natives Responses		
	Most Frequent	**Second Most Frequent**
GSCG	Backup, 27.7%	Find my phone, 20.3%
MSCG	None, 31.7%	Backup, 29.5%
HSCG	Backup, 57.1%	Find my phone, 57.1%
Security Experts Estimation		
GSCG	I do not know these tools, 50.0%	Find my phone, 29.4%
MSCG	Backup, 29.4%	Find my phone, 29.4%
HSCG	Find my phone, 26.5%	Remote wipe, 23.5%

Responses for "None"

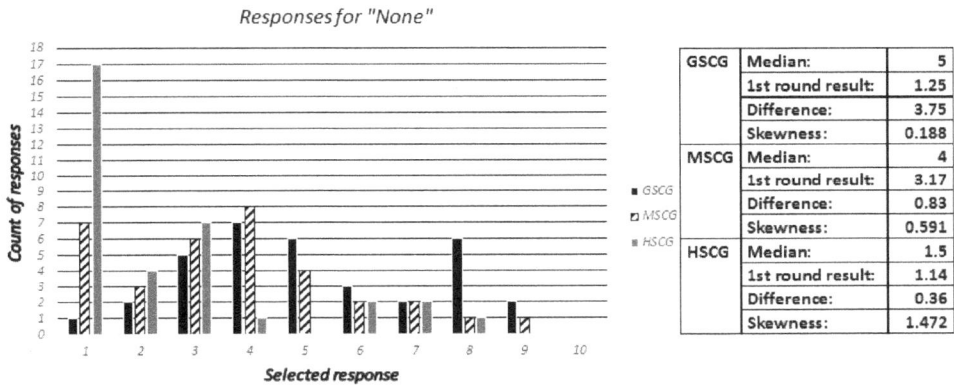

GSCG	Median:	5
	1st round result:	1.25
	Difference:	3.75
	Skewness:	0.188
MSCG	Median:	4
	1st round result:	3.17
	Difference:	0.83
	Skewness:	0.591
HSCG	Median:	1.5
	1st round result:	1.14
	Difference:	0.36
	Skewness:	1.472

Figure 15. Expert predictions for question 11.

Responses for "I do not know these tools"

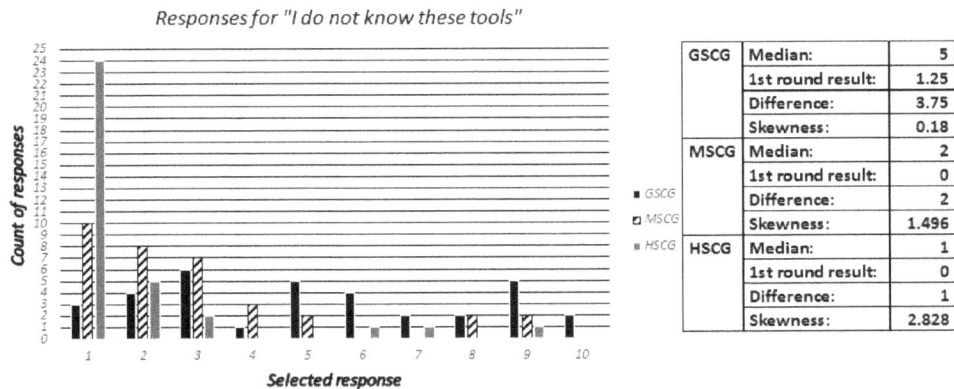

GSCG	Median:	5
	1st round result:	1.25
	Difference:	3.75
	Skewness:	0.18
MSCG	Median:	2
	1st round result:	0
	Difference:	2
	Skewness:	1.496
HSCG	Median:	1
	1st round result:	0
	Difference:	1
	Skewness:	2.828

Figure 16. Expert predictions for question 11, "None".

4.4.3. Question 12

During the first stage of our study, this section concluded with a set of questions towards the digital natives about their choices in respect to the selection and use of passwords. Accordingly, the security experts have been asked to estimate how many out of ten digital natives across all SCGroups have reported to:

- Always use the same password,

- Use small variations of the same password for different applications,
- Always use different passwords.

A summary of the results for this set of questions is presented in Figures 17–19.

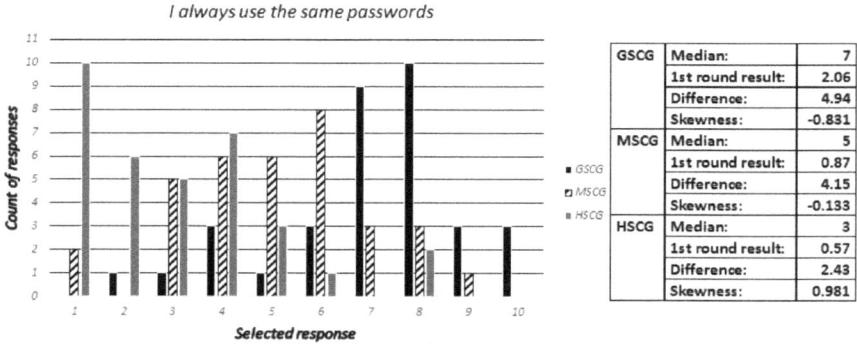

Figure 17. Expert predictions for question 12.

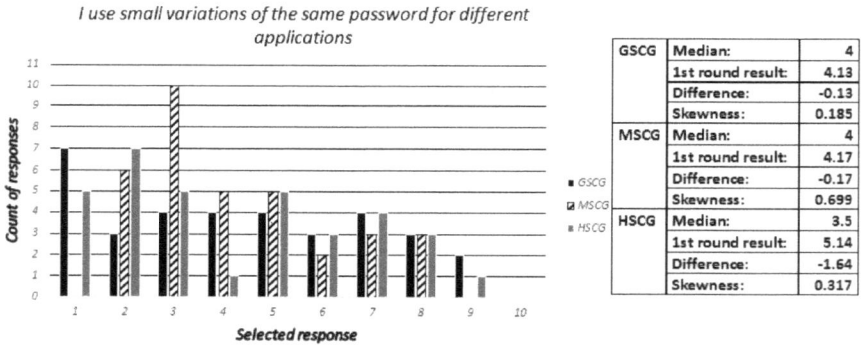

Figure 18. Expert predictions for question 12.

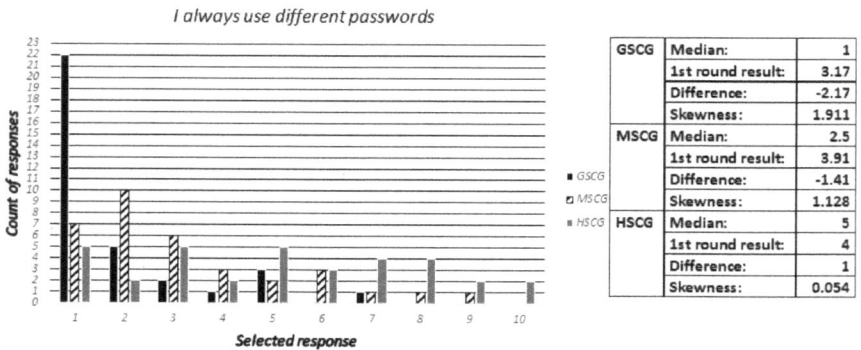

Figure 19. Expert predictions for question 12.

5. Summary of Results and Discussion

Table 4 summarizes the expert prediction accuracy for each area and the total. Starting with the use of mobile devices, our results show that the Experts only estimate correctly 2 out of 10 possible for GSCG. The low prediction rate for the GSCG continues throughout the results, in which the experts missed all three on the connectivity and network access parts. They got 1 of 4 right on the Management of credentials part, and 4 out of 10 regarding protection mechanisms. The total prediction accuracy of the experts regarding the GSCG was 26%, which indicates a poor understanding of the group for all areas.

Table 4. Summary of prediction accuracy for the included findings.

Area	Amount	GSCG Correct	Missed	MSCG Correct	Missed	HSCG Correct	Missed	Total Accuracy
Use of Mobile Device	10 (HSCG 12)	2	8	5	5	7	5	43.75%
Connectivity and Network Access	3	0	3	2	1	2	1	44.44%
Management of credentials	4	1	3	0	4	2	2	25%
Protection Mechanisms	10	4	6	4	6	6	4	46.67%
Total for all	27 (HSCG 29)	7	20	11	16	17	12	42.16%
Total Accuracy		25.92%		40.70%		58.62%		

The expert predictions regarding the MSCG was 15% better than GSCG. This result was in particular caused by the predictions regarding the use of mobile devices and connectivity where the experts got 50% and 75% of the predictions right. The predictions for the HSCG were 18% better than the MSCG, which brings forth a clear trend: the experts are better calibrated to predict users with better security knowledge. However, the total accuracy of 59% for expert predictions regarding the HSCG is not a very high number. Considering the total average accuracy, the results for three areas, use of mobile devices, connectivity and network access, and protection mechanisms) are within the range 44%–47%. Management of credentials is the outlier with only 25%. The results indicate an increased expert understanding of user behavior with the level of security knowledge. However, the total expert prediction accuracy for all three groups was 42%, which indicates a poor understanding of the digital natives.

The results of this study present some notable findings, with respect to how security experts perceive the digital natives as users of mobile devices. From questions 10 and 11, we see that security professionals have a relatively good understanding of the technology penetration rates for various security related tools such as access control methods.

Furthermore, the results across the survey allow us to identify that security experts consistently underestimate the security awareness of the general population, represented by the GSCGroup. This is noticeable across all the questions, with minor exceptions, such as Question-3, and some sub-categories such as Question-4b ("I always check the required permissions before installing an application"), Question-9b ("I always log-out after finishing using an application"), and Question 12-b ("I use small variations of the same password for different applications"). Moreover, this consistent deviation between the median value of the experts' responses and the first round of results is noticeably extended, with ten questions providing a difference of more than 3 (30%) and up to 5.5 (55%).

The results are different in respect to the HSCG group, which consists of university students in the field of information security. For this group, the responses for seventeen questions have been estimated correctly by the experts, while, in five cases, the security awareness of this group has been overestimated, and only in one case has been underestimated. Furthermore, it is noticeable that the distribution of the responses for a sub-set of questions does not present a clearly defined tendency or concentration around a central value. From this, we can extract that the user-models of the security professionals include randomness, and may be biased by personal perception. Examples of this are noticeable in Question-1-GSCG, Question-3, Question-5-HSCG, and Question-8-MSCG.

6. Limitations & Future Work

We conducted this study under the limitation that a prior study collected the GSCG dataset. Therefore, access to the raw data was not possible, which restricted the possibility for deeper statistical analysis. The diversities in both national and cultural backgrounds we find at universities make them ideal for this type of study. However, although each data sample (GSCG, MSCG, and HSCG) was collected at universities, and it is likely that the majority of the respondents originate from the country where the university is situated and that there is a culture bias in our samples. Furthermore, the two samples for the HSCG and the experts were both small (34 and 35 respondents), which makes them vulnerable to outliers. However, the results we provide in this study provide a strong incentive for future studies within user modeling of the digital natives. A path for future research is validation studies where the researchers go deeper into each area to determine more precisely where the expert understanding is poor. Based on the proposed studies, researchers can create better training programs to improve the understanding of the digital natives, which will lead to enhanced security solutions for the new generation.

7. Conclusions

Digital natives make extensive use of mobile devices, while such devices are increasingly integrated into complex socio-technical systems with critical security implications. In our study, we sought to identify how user behavior can affect the security of emerging mobile technologies. Therefore, at the initial stage, a survey allowed the extraction of findings with respect to how the digital natives use their mobile devices and perceive associated risks according to their background and level of security awareness. Accordingly, in this article, we approached this topic from the perspective of experts, who are involved in the design, operational support, and analysis of such systems. Our results suggest that the experts understanding over user behavior does not follow a solidified user-model, especially on the general population. Furthermore, in some of the identified topics, influences from personal perceptions and randomness was noticeable. Accordingly, improving security over such systems would require not only the enhancement of the users' security awareness, but also the improvement of existing user-models and their dissemination within the design and analysis phases.

Author Contributions: The authors contributed equally to this work.

Conflicts of Interest: The authors declare no conflict of interest.

References

1. Felt, A.P.; Ha, E.; Egelman, S.; Haney, A.; Chin, E.; Wagner, D. Android permissions: User attention, comprehension, and behavior. In Proceedings of the Eighth Symposium on Usable Privacy and Security, Washington, DC, USA, 11–13 July 2012; p. 3.
2. Kelley, P.G.; Consolvo, S.; Cranor, L.F.; Jung, J.; Sadeh, N.; Wetherall, D. A conundrum of permissions: Installing applications on an android smartphone. In Proceedings of the International Conference on Financial Cryptography and Data Security, Kralendijk, Bonaire, 27 February–1 March 2012; pp. 68–79.
3. Ophoff, J.; Robinson, M. Exploring end-user smartphone security awareness within a South African context. In Proceedings of the Information Security for South Africa, Johannesburg, South Africa, 13–14 August 2014; pp. 1–7.

4. Gkioulos, V.; Wangen, G.; Katsikas, S.K.; Kavallieratos, G.; Kotzanikolaou, P. Security Awareness of the Digital Natives. *Information* **2017**, *8*, 42.

5. Lella, A.; Lipsman, A. The US Mobile App Report. 2014. Available online: http://www.comscore.com/Insights/Presentationsand-Whitepapers/2014/The-US-Mobile-App-Report (accessed on 8 April 2015).

6. Prensky, M. Digital natives, digital immigrants part 1. *Horizon* **2001**, *9*, 1–6.

7. Bennett, S.; Maton, K.; Kervin, L. The 'digital natives' debate: A critical review of the evidence. *Br. J. Educ. Technol.* **2008**, *39*, 775–786.

8. Chin, E.; Felt, A.P.; Sekar, V.; Wagner, D. Measuring user confidence in smartphone security and privacy. In Proceedings of the Eighth Symposium on Usable Privacy and Security, Washington, DC, USA, 11–13 July 2012; p. 1.

9. Mylonas, A.; Gritzalis, D.; Tsoumas, B.; Apostolopoulos, T. A qualitative metrics vector for the awareness of smartphone security users. In Proceedings of the 10th International Conference on Trust, Privacy and Security in Digital Business, Prague, Czech, 28–29 August 2013; pp. 173–184.

10. Mylonas, A.; Kastania, A.; Gritzalis, D. Delegate the smartphone user? Security awareness in smartphone platforms. *Comput. Secur.* **2013**, *34*, 47–66.

11. Parker, F.; Ophoff, J.; Van Belle, J.P.; Karia, R. Security awareness and adoption of security controls by smartphone users. In Proceedings of the Second International Conference on Information Security and Cyber Forensics (InfoSec), Cape Town, South Africa, 15–17 November 2015; pp. 99–104.

12. Markelj, B.; Bernik, I. Safe use of mobile devices arises from knowing the threats. *J. Inf. Secur. Appl.* **2015**, *20*, 84–89.

13. Markelj, B.; Zgaga, S. Comprehension of cyber threats and their consequences in Slovenia. *Comput. Law Secur. Rev.* **2016**, *32*, 513–525.

14. Kandias, M.; Mylonas, A.; Virvilis, N.; Theoharidou, M.; Gritzalis, D. An insider threat prediction model. In Proceedings of the International Conference on Trust, Privacy and Security in Digital Business, Bilbao, Spain, 30–31 August 2010; pp. 26–37.

15. Workman, M.; Bommer, W.H.; Straub, D. Security lapses and the omission of information security measures: A threat control model and empirical test. *Comput. Hum. Behav.* **2008**, *24*, 2799–2816.

16. Bulgurcu, B.; Cavusoglu, H.; Benbasat, I. Information security policy compliance: An empirical study of rationality-based beliefs and information security awareness. *MIS Q.* **2010**, *34*, 523–548.

17. Ng, B.Y.; Kankanhalli, A.; Xu, Y.C. Studying users' computer security behavior: A health belief perspective. *Decis. Support Syst.* **2009**, *46*, 815–825.

18. Asgharpour, F.; Liu, D.; Camp, L.J. Mental models of security risks. In Proceedings of the 11th International Conference on Financial Cryptography and Data Security, Scarborough, Trinidad and Tobago, 12–16 February 2007; pp. 367–377.

19. Camp, L.J. Mental models of privacy and security. *IEEE Technol. Soc. Mag.* **2009**, *28*, doi:10.1109/MTS.2009.934142.

20. Herley, C. So long, and no thanks for the externalities: The rational rejection of security advice by users. In Proceedings of the Workshop on New Security Paradigms Workshop, Oxford, UK, 8–11 September 2009; pp. 133–144.

21. Ariu, D.; Bosco, F.; Ferraris, V.; Perri, P.; Spolti, G.; Stirparo, P.; Vaciago, G.; Zanero, S. Security of the Digital Natives. *Available SSRN* **2014**.

22. Norman, G. Likert scales, levels of measurement and the "laws" of statistics. *Adv. Health Sci. Educ.* **2010**, *15*, 625–632.

future internet

MDPI

Article

Participation and Privacy Perception in Virtual Environments: The Role of Sense of Community, Culture and Gender between Italian and Turkish

Andrea Guazzini [1,2,*], Ayça Saraç [3], Camillo Donati [1], Annalisa Nardi [1], Daniele Vilone [4,5] and Patrizia Meringolo [1]

[1] Department of Education and Psychology, University of Florence, Via di San Salvi 12, 50135 Firenze, Italy; camillodonati86@gmail.com (C.D.); annina.nardi@gmail.com (A.N.); patrizia.meringolo@unifi.it (P.M.)
[2] Center for the Study of Complex Dynamics (CSDC), University of Florence, Via di San Salvi 12, 50135 Firenze, Italy
[3] Department of Educational Sciences, Çukurova University, Balcali, Adana 01330, Turkey; asarac@cu.edu.tr
[4] LABSS (Laboratory of Agent Based Social Simulation), Institute of Cognitive Science and Technology, National Research Council (CNR), Via Palestro 32, 00185 Rome, Italy; daniele.vilone@gmail.com
[5] Grupo Interdisciplinar de Sistemas Complejos (GISC), Departamento de Matemáticas, Universidad Carlos III de Madrid, 28911 Leganés, Spain
* Correspondence: andrea.guazzini@unifi.it; Tel.: +39-055-2756134

Academic Editor: Georgios Kambourakis
Received: 18 October 2016 ; Accepted: 4 April 2017; Published: 7 April 2017

Abstract: Advancements in information and communication technologies have enhanced our possibilities to communicate worldwide, eliminating borders and making it possible to interact with people coming from other cultures like never happened before. Such powerful tools have brought us to reconsider our concept of privacy and social involvement in order to make them fit into this wider environment. It is possible to claim that the information and communication technologies (ICT) revolution is changing our world and is having a core role as a mediating factor for social movements (e.g., Arab spring) and political decisions (e.g., Brexit), shaping the world in a faster and shared brand new way. It is then interesting to explore how the perception of this brand new environment (in terms of social engagement, privacy perception and sense of belonging to a community) differs even in similar cultures separated by recent historical reasons. Recent historical events may in effect have shaped a different psychological representation of Participation, Privacy and Sense of Community in ICT environments,determining a different perception of affordances and concerns of these complex behaviors. The aim of this research is to examine the relation between the constructs of Sense of Community, Participation and Privacy compared with culture and gender, considering the changes that have occurred in the last few years with the introduction of the web environment. A questionnaire, including ad hoc created scales for Participation and Privacy, have been administered to 180 participants from Turkey and Italy. In order to highlight the cultural differences in the perception of these two constructs, we have provided a semantic differential to both sub-samples showing interesting outcomes. The results are then discussed while taking into account the recent history of both countries in terms of the widespread of new technologies, political actions and protest movements.

Keywords: virtual environment; participation; privacy; sense of community; cross-cultural psychology; social psychology

1. Introduction

The social nature of human beings has attracted the interest of researchers since the beginning of science. Being part of a social entity and actively participating in it is a part of all our lives. This natural tendency explained for the first time by Social Darwinism represents one of the environmental pressures that has probably affected most of human evolution. Cooperation, competition and other social skills (e.g., social problem solving, social self-efficacy) as well as dynamics (e.g., individualism and pluralism, in-group favoritism) have determined human evolution and are still distinguishing our society more than other environmental features. Within such complex dynamics, social and cultural groups have developed different strategies in order to maintain an efficient way to cooperate within their community while simultaneously adopting an effective strategy to compete with the other groups within their ecological niche.

Nowadays the revolution in information and communication technologies (ICT) is changing the old social equilibrium of the world by providing a brand new way to interact with others, new social structures and possibilities, as well as new "ecological niches" to be explored and exploited.

In the 1990s, Dunbar's number theory set the cognitive limit of the number of people with whom one can maintain stable social relationships to 150 [1,2]. It is clear that with the massive use of social networks and online communication tools the number of social relations we deal with daily is increasing, which suggests the need to re-think the concepts of ecological niches as well as the concepts of community, participation and privacy. Similarly, even the ecological system theory [3] needs to be updated by considering the possible relations that can be developed through new scenarios such as the world wide web. Such a brand new way to interact is in fact affecting our lives as well as the world organization, and presumably scientists should consider this as the most important environmental factor that will affect the human evolution in the future. For example recent and important historical events (i.e., Arab Spring, Occupy Wall Street Movements, anti-coup movements in Turkey) took place thanks to the possibility provided by social networks to gather an extraordinary amount of people in such a brief amount of time: before the ICT revolution it would have been at least harder to make those movements happen. Sure enough social networks, and new media in general, are having a key role in driving debates and public opinion in consequence of socially relevant events and can be seen as mediators of the consequences of these. At the same time recent historical events are shaping our way to interact with ICT tools so that similar cultures are approaching social networks and new media in a different way, depending on the degree of affordances and concerns related to their use.

In order to properly approach this environmental change, we need to consider the cultural differences characterizing the interaction between people within this brand new world (i.e., perception of privacy, virtual sense of community and web based participation) and distinguishing the new equilibria that interconnects human society through the world wide web [4]. Understanding the human factors involved in the use of social networks and online applications, such as mobile applications or crowdsourcing platforms, is now one the most challenging objectives for several disciplines [5,6]. In fact by considering the massive and daily use of web tools in our lives, we need to understand how several well-known factors and socio-psychological dynamics in literature can be translated into virtual contexts. The opportunities in terms of communication and the sharing of knowledge that the "virtual world" is providing us are astonishing. However in order to gain an actual comprehension of the usage of such tools, a deeper understanding of the complex dynamics behind those processes needs to be achieved. While a proper and comprehensive model for the human interaction with technologies is still under development, it can be useful to approach this research by using well known techniques adopted in the past in the "real world" as the cross-cultural analysis of features involved in the behaviors of interest.

Cross-cultural differences involved in the perception of behaviors, attitudes or emotions have frequently been a precious tool in order to understand peculiarities or commonalities in several areas of research . The enrichment provided by different cultures can, in fact, highlight peculiarities related to the context or the history of a population that may help researchers in modeling complex behaviors.

Investigating the connections between well known constructs and taking into account the differences due to cultural specificity or recent historical events is necessary for researchers to understand new important features, excluding some others related to a specific context, but showing the important factors in common with the human race.

The main objective of our study is to measure the relationship between Sense of Community, Participation and Privacy compared to culture and gender, taking into account two cultures (Italian and Turkish) similar for many features, but surely different for recent historical events. After a brief theoretical introduction to the constructs (Sense of Community, Participation and Privacy), we will describe our study in which 180 digital native participants (83 Italians and 97 Turkish People) have been asked to fill in an online questionnaire.

1.1. Sense of Community

Sense of community (SoC) is a construct largely used in community psychology, where it has been used to describe the complex relationship that an individual has while feeling as an active part of his community. Since the term "community" has a multi-level nature, which according to Brodsky and Marx (2002) can be referred to neighborhood, block groups, housing complexes, schools and cities, it has been studied in a large variety of contexts [7,8]. Given that several definitions have been taken into account, the variety of features that this construct can express for different individuals in different settings is phenomenal. Sarason defined it as the feeling that community members have about each other [9,10], McMillan and Chavis (1986) claimed that SoC is a concept related to membership, emotional safety, identity, belonging and attachment to a group [11]. The perception of similarity to others and willingness to communicate with people are also key features in defining SoC [12]. According to these authors, when we feel part of a community we are more willing to share responsibility, to improve face to face relationships and to participate with others in social activities. On the other hand, several social scientists have shown that the increased complexity and change in technologies have affected the meaning and importance of communities [7]. Relevant to this new approach to SoC, in which the contexts are evolving from a physical reality into a virtual scenario, all research is aimed at identifying a definition of the Sense of Virtual Community [13–15]. The Sense of Virtual Communities has been found to be overlapping in the SoC in many features: for example participants in some studies by Francescato (2006; 2007) have shown similar levels in collaborative learning, developed self-efficacy, developed problem solving and perception of social capital when comparing face to face and online groups [15,16].

1.2. Participation

Due to its multi-disciplinary nature, participation has been investigated through approaches coming from several academic disciplines (e.g., Sociology, Anthropology Psychology) even though a proper descriptive and comprehensive model is still lacking at the moment. Given that, it can be analyzed starting from different levels: macro-social/institutional level, micro-social levels, psychological level. According to Cicognani and Zani (2011), when discussing participation on the macro-social/institutional level we refer to the electoral system, civic education, culture, religion, social-economic development and the history of the country in question [17]. On the micro-social level, contexts such as family, friends, school, and voluntary associations are taken into account . On the psychological level, Wilkenfeld et al. (2010) highlight the importance of participation, on both civic and political issues as well as on the cognitive and social development of adolescents [18]. A key feature in promoting participation is the presence of a collaborative group sustaining the action. Sense of collective identity stimulates engagement in collective actions [19] as well as identification with the group [17]. To this effect, SoC [11] has been proven to have a positive, bidirectional, association with participation [20]. One key feature related to participation is also the construct of empowerment [21] and as found in Zimmermann (2000) the action-oriented nature of empowerment finds a natural consequence in both individual and social participation [22]. Combining the contribution from psychological literature and the recent findings coming from the renewed interest in this construct, especially from Information and Communication Technologies studies, several features have been taken into account to find the proper incentives promoting participation (inhibiting competition). For instance Gachter investigated the role of rewards and punishment, finding out that punishments can increase cooperation even in the long term, because the gains from cooperation itself give enough gains to became stable even if the punishment is removed [23]. In a more recent review on this topic, Gachter shows that even the size of the incentives has a crucial role in these dynamics. In a contest of rewards, a very large incentive can enhance a stronger cooperation, while with a higher punishment a low threat can be enough to reach the same result [24–26]

1.3. Privacy

With the increasing use of social networks (e.g., Facebook, Twitter, Google+), and collaborative web platforms (i.e., crowdsourcing tools) people find the opportunity to share ideas and feelings with others on the Internet every day [27]. The ICT revolution, the new media and the Internet, improved deeply the relevance of privacy, as well as of security requirements [28–30]. The same concept of Privacy has been reformulated and represented as a "multidimensional object", no longer representable using the classic definitions [31]. While exploring these brand new and innovative media, we learned that we also need to take into account the potential dangers that can occur while exchanging personal data online. Participating in social media is highly related to the feeling of safety that people experience while sharing their contents on the web. According to Song, Hao and Daqing (2013), people that have a higher trust in social networks are more likely to participate actively than those who have less [32]. When people feel a risk about sharing personal data on social networks they do not contribute to them. As mentioned above in SoC, behavior in real physical contexts and web scenarios seem to be overlapping in some of its features. In fact, privacy is a very important dimension of human life, affecting personal and social lives. Since it involves the control of the amount of contact with others and the perceived safety of these interactions, Solove (2008) has found that when people do not meet their privacy needs, it may result in antisocial and stressful behaviors [33]. Altman has defined privacy as "the selective control of access to information to the self or to one's group" [34]. According to Pedersen (1997) six types of privacy can be identified: solitude, isolation, anonymity, reserve, intimacy with friends and family [35]. Privacy attitudes have been measured in a sample of 210 men and 165 women from high schools and colleges in Turkey [36]. Results show that women have higher mean scores for measures of intimacy with friends and lower mean scores for isolation and reserve than men. There were no mean differences for solitude, intimacy with family, and anonymity. Cross-cultural differences in the perception of privacy regulation have been investigated by Kaya and Weber in 2003, involving American and Turkish students [37]. The result showed that American students desired more privacy in their residence hall rooms than Turkish students. Regardless of culture, males reported a greater desire for privacy than females. The relationship between privacy and security was investigated in a study regarding the educational use of cloud service such as social networks, Google drive and Dropbox [38]. Here, security was taken into account as the degree to which students believed that cloud-services were secure platforms for storing and sharing sensitive personal data. The results showed that the perception of a low level of security may have affected students' attitudes towards using such services. In other words, students with low tolerance for technological risks may defer their use of these services and privacy concerns may then impede attitude towards educational use of cloud services. Several scales have also been recently developed in order to investigate privacy.

1.4. Aims and Hypotheses of the Present Study

The main objective of our study is to measure the relationship among the constructs of Sense of Community, Participation, Privacy according to culture and gender. Therefore, we examined the score on the scale for Sense of Community, Participation and Privacy in respect to gender and culture. The main hypothesis of the research is that social movements and uprisings, characterizing the recent history of Turkey, are affecting the perception of the web based participation and privacy psychological constructs, beyond the structural differences of the two populations considered in the study. Taking into account two cultures (Italian and Turkish) similar for many features, but surely different for recent historical events, we wanted to highlight the effect of these occurrences on the usage of social media. In particular, we expect that Turkish people may have a higher perception of the importance of participating and considering the consequences of privacy in virtual environments as a consequence of the important role that social media are having on their recent history. We also wanted to assess the influence of culture and gender on the perceived Sense of Community, Sense of Confidence/Concern toward the representation of the construct of Participation and Privacy (in order to consider cultural values referred to this construct, we provided an Italian/Turkish definition of the construct). In order to appreciate more how cultural differences are involved in representing the constructs of Participation and Privacy, we provided a semantic differential [39] for both Italian and Turkish sub-samples. The semantic differential is a rating scale in which respondents are asked to choose their position on a continuum between two polar adjectives. This measure has also been analyzed in respect to gender. At the end of our analysis, we present the correlation structure between all the measures adopted for both the entire sample as well as the single sub-samples.

1.5. Methods

1.5.1. Sampling and Participants

A total of 180 participants were recruited on a voluntary basis for the online questionnaires of our study. Every participant received an e-mail on their University account containing a link to the online questionnaires and a brief explanation on the research. We divided the sample on the basis of their origin (i.e., Italy/Turkey). The Italian sub-sample (29 M; 54F) shows a mean age of 24.69 S.D. 3.9 (Male = 24.24 S.D. 4.46; Female = 24.93 S.D. 3.57) and the Turkish sub-sample (49 M; 48 F) shows a mean age of 22.4 S.D. 2.67 (Male = 22.96 S.D. 2.68; Female = 21.83 S.D. 2.55). Most of the participants were students (84%) recruited from The University of Florence (Italian sub-sample) and the Cukurova University (Turkish sub-sample). In Table 1 we report the sample size estimation [40] to ensure a correct evaluation of the test.

Table 1. Sample size estimation to compare 2 means from 2 samples with a 2 sided equality hypotheses [40], with a samples ratio $K = \frac{N_a}{N_b} = 0,85$, and an actual sample size $N_b = 97$.

Dimension		Reference Sample Mean (SD)	Test Sample Mean (SD)	Power $1 - \beta$	Type I Error α	Required Sample Size
Privacy	Affordance	12.5 (2.3)	14.7 (3.2)	0.9	0.01	69
	Concerns	14.5 (2.9)	17.5 (4.4)	0.9	0.01	70
Participation	Affordance	16.1 (3.7)	18.6 (3.9)	0.9	0.01	77
	Concerns	12.6 (3.6)	14.8 (2.1)	0.9	0.01	87
Sense of Community		53.8 (9.5)	50.3 (11.3)	0.8	0.1	89

1.5.2. Procedures

The main objective of this study, as mentioned above, is to measure the relationship between Sense of Community, Participation and Privacy, in two different cultures (Italian and Turkish). In order to assess this relationship in respect to Culture and Gender, a set of questionnaires was prepared.

The scale we used to assess Sense of Community was the union of two different scales: The Classroom and Community Inventory (CSCI) as found in Rovai, Wighting and Lucking (2004), and the Sense of Community in School as found in Vieno, Santinello, Pastore and Perkins (2007), for a total of 16 items on a 5 point Liker scale, from 1 = "strongly" disagree to 5 = "Strongly agree" [41,42]. Since this questionnaires are aimed to assess Sense of Community in a scholastic environment, we took in consideration only participants that at the moment of the administration were university students or university workers (e.g., teachers, researchers, PhD students).

In order to obtain a measurement for both the Participation and Privacy constructs, two different questionnaires were developed. Both questionnaires were structured following this scheme: definition, semantic differential, and scale.

Definition: for both dimensions a general definition of the construct has been provided using those found in popular dictionaries (Treccani dictionary for Italian, "Wikipedia" and "Türk Dil Kurumu" for Turkish) for both Italian and Turkish sub-samples.

Semantic differential: the Semantic differential technique was used in order to appreciate more the cultural differences regarding both constructs [39]. According to the Osgood's semantic differential (1964) a list of 10 items made of two bipolar adjectives was provided (e.g., warm-cold; useful-useless; safe-dangerous). Participants were asked to choose where his/her position lie in the continuum (a Likert Scale from 1 to 10) from the positive to the negative side (adjective) of each of the 10 items.

Scale: for both Participation and Privacy an ad-hoc questionnaire was developed. Both scales presented 10 items assessing participants' perception of Participation/Privacy in relation to different contexts (real life, online and social networks). Items were alternated into two sub-sections: 5 items assessed the confidence towards these constructs, 5 assessed concerns towards them. Answers on a 5-point Likert scale (1 = "Absolutely no", 2 = "A few", 3 = "Moderately", 4 = "A lot", 5 = "Very much") were coded and summed up in order to obtain a general score for both Concerns/Confidence sub-scales.

1.5.3. Data Analysis

The data analysis activity was structured in two main phases. In the first one, we calculated the descriptive statistics, assessing the pre-conditions required by the subsequent inferential analysis. In particular, we checked the Gaussian distribution of the continuous variables, i.e., skewness and kurtosis $\in (-1; +1)$, and the sufficient balancing and size of the sub-samples of interest (i.e., gender and nationality). Then in the second phase we conducted the inferential analyses required by our main hypothesis. In particular, within the context of this research the Pearson's r correlation between the variables such as participation, privacy and sense of community were analyzed. Furthermore, an ANOVA analysis was conducted to find out if these three variables differentiate in terms of nationality and gender factors. SPSS.20 was used for these analyses.

2. Results

In this section, the main results of our research are presented. Descriptive statistics are presented first, where we show the results of the socio-demographic and study variables (Participation, Privacy, Sense of Community) of our sample. In the following section, we show results of the inferential analysis (i.e., correlation structures, ANOVA analysis, effects of socio-demographics on a semantic differential scale).

2.1. Descriptive Statistics

2.1.1. Sociodemographic Variables

The task has been administered to Italian and Turkish university students. The sociodemographic analyses of the experimental sample are reported in Tables 2 and 3. The Italian sub-samples shows a mean age of 24.68 and is composed of 54 females and 29 males, the Turkish (mean age = 22.40) of 48 females and 49 males.

Most of the participants were students (84%), single (75%) and had an high degree of education (Bachelor degree = 71%). The Turkish sub-sample seem to be more uniform regarding civil status and education Table 3.

Table 2. Sizes and average of ages for the Italian and Turkish sub-samples are reported. The standard deviation of each variable (σ) is reported in brackets.

Subsample	n	Age
Italian	83	24.68 (3.9)
Turkish	97	22.40 (2.7)
It Females	54	24.92 (3.6)
It Males	29	24.24 (4.5)
Tk Females	48	21.83 (2.6)
Tk Males	49	22.96 (2.7)

Table 3. Summary of descriptive statistics about the discrete sociodemographic variables of the sample.

Variable	Total	Italian Subsamples		Turkish Subsamples	
		Female	Male	Female	Male
Occupation	180				
Student	152	43	20	44	45
Employed	20	4	8	4	4
Unemployed	8	7	1	0	0
Civil status	180				
Single	136	24	19	47	46
Engaged	38	29	6	1	2
Married	6	1	4	0	1
Education	180				
High Sc.	24	12	11	0	1
Bachelor	128	27	11	44	46
Mast Sc.	28	15	7	4	2

2.1.2. Study Variables Results

Our scales about Participation and Privacy seem to be able to discern between the confidence and the concerns the participant has about these two dimensions (Table 4). In the Italian sample, we found general lower values compared to the Turkish sub-sample, for both Confidence and Concerns. Average scores suggest that Italian females are more confident (16.4) than males (15.4) and less concerned (female = 12.9; male = 12.1) about participation. On the contrary they are more concerned (14.9) than males (13.9) and less confident (female = 12.1; male = 13.3) about privacy. In the Turkish sample we see the same trend, but with higher values. Males and females have similar values about Participation Confidence (18.9 and 18.3), generally higher than the Concerns (female = 14.6; males = 15.1). Like in the Italian sample, Turkish participants have higher values in Privacy Concerns (18.1 for female and 16.9 for male) than in Privacy Confidence (13.6 and 15.7) and females show higher values than males on concerns for both measures. Italian females (55.17) show higher values on the sense of community scale than Italian males (51.38). The Turkish sub-sample has generally lower values than the Italians, with similar scores for females (50.23) and males (50.44).

According to the Semantic differential, the most interesting results about Participation is the elevated contrast between Italians and Turkish People. As you can see in Figure 1 Italians assessed almost all the adjectives at one of the extremes of the continuum; both males and females, showed very high scores for "Useless-Useful", "Bad-Good", "Dirty-Clean" and "Personal-Social"; for almost all other differentials the scores are very low, except for "Easy-Di cult". Turkish samples present a totally different trend, where all values are low, only "Personal-Social", "Easy-Di cult" and "Concrete-Abstract" have slightly higher values. For both Italians and Turkish people, the differences between males and females are minimal.

Table 4. In table the average and standard deviation (in brackets) regarding Participation, Privacy and Sense of Community scores are reported for each subsample.

Measure	Italian Subsamples		Turkish Subsamples	
	Female	**Male**	**Female**	**Male**
Participation				
Confidence	16.4 (3.3)	15.4 (4.3)	18.3 (3.9)	18.9 (4.0)
Concerns	12.9 (3.6)	12.1 (3.9)	14.6 (2.6)	15.1 (3.4)
Privacy				
Confidence	12.1 (1.8)	13.3 (3.0)	13.6 (3.2)	15.7 (3.0)
Concerns	14.9 (2.9)	13.9 (2.8)	18.1 (4.2)	16.9 (4.5)
Sense of Community	55.17 (7.4)	51.38 (12.3)	50.23 (11.6)	50.44 (11.1)

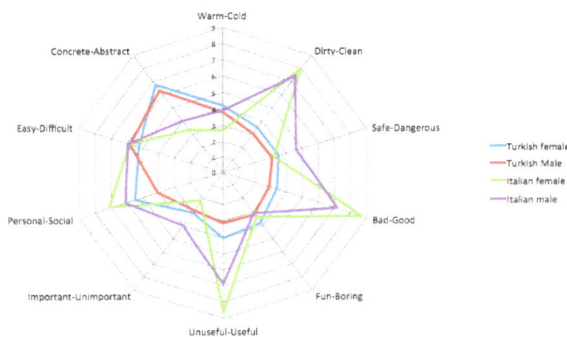

Figure 1. Semantic differential for all the sub-samples with respect to the concept of "Participation".

In the Privacy Semantic Differential Figure 2 the Turkish sample shows a trend similar to the Participation one with mid-low values in almost all items, except for "Personal-Social", "Easy-Difficult" and "Concrete-Abstract". The Italian sample has some similarities, but we can see lower values for "Important-Unimportant", whereas we found mid values in the "Easy-Difficult", "Concrete-Abstract" and "Warm-Cold" items. Exactly like the Participation Semantic differential and even for privacy, the differences between males and females are very low in both samples.

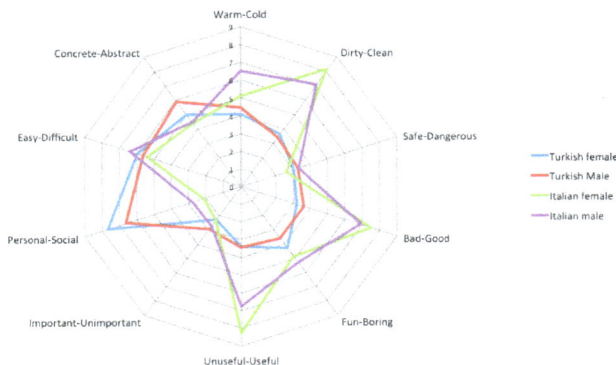

Figure 2. Semantic differential for all the sub-samples with respect to the concept of "Privacy".

Finally, the sense of community scale reports higher values for Italians in general. In particular, females report higher values in the Italian sub-sample (55.17 vs. 51.38), and lower but similar values in the Turkish sub-sample (50.23 vs. 50.44), as indicated in Table 4.

2.2. Inferential Statistics

The inferential analysis was structured in three different phases investigating the correlation structure characterizing the observable quantities taken into account in this study (i.e., age, education, sense of community, participation and privacy). The second phase ran three ANOVA analyses in order to investigate the univariate and combined effects of nationality and gender on the three fundamental variables defining the constructs under scrutiny (i.e., sense of community, participation and privacy). Finally, the last phase was dedicated to assessing the effects of the socio-demographic factors of interest (i.e., nationality and gender) on the differential semantic dimensions describing the participation and privacy perception.

2.2.1. Univariate Analysis: Correlation Between the Order Parameters

The Pearson's r correlation statistic was adopted to estimate the relations between the fundamental variables of interest with respect to the 4 sub population considered by our study Table 5.

Table 5. Correlations between the dependent measures.

Measure	Age	Education	Sense of Community	Participation Submeasures		Privacy Submeasures	
				Confidence	Concerns	Confidence	Concerns
Sense of Community							
It. Female	ns	0.28 *	1	ns	ns	ns	ns
It. Male	ns	−0.45 *	1	0.69 **	0.54 **	ns	0.58 **
Tk. Female	ns	ns	1	ns	ns	ns	ns
Tk. Male	ns	ns	1	0.39 **	−0.33 **	ns	ns
Par. Confidence							
It. Female	ns	ns	ns	1	−0.53 **	0.28 *	ns
It. Male	ns	ns	0.69 **	1	ns	ns	0.40 *
Tk. Female	ns	ns	ns	1	ns	0.61 **	0.35 *
Tk. Male	ns	ns	0.39 **	1	ns	0.40 **	ns
Par. Concerns							
It. Female	ns	ns	ns	−0.53 **	1	ns	0.34 *
It. Male	ns	ns	0.54 **	ns	1	ns	0.47 **
Tk. Female	ns	ns	ns	ns	1	ns	0.29 *
Tk. Male	ns	ns	−0.33 *	ns	1	ns	0.40 **
Pri. Confidence							
It. Female	ns	−0.31 *	ns	0.28 *	ns	1	ns
It. Male	ns	ns	ns	ns	ns	1	ns
Tk. Female	0.29 *	ns	ns	0.61 **	ns	1	ns
Tk. Male	ns	ns	ns	0.40 **	ns	1	ns
Pri. Concerns							
It. Female	ns	−0.40 **	ns	ns	0.34 *	ns	1
It. Male	ns	ns	0.58 **	0.40 *	0.47 **	ns	1
Tk. Female	ns	ns	ns	0.35 *	0.29 *	ns	1
Tk. Male	ns	ns	ns	ns	0.40 **	ns	1

Pearson r correlation, *: $p < 0.05$, **: $p < 0.01$, ns: not significant.

The age of the participants appears to have no significant effects. The only exception is related to the Turkish female sample where a weak relation emerged between age and privacy confidence.

Education appears to be related to the sense of community only in the Italian sample. In particular, it showed quite an interesting reversed effect regarding gender for Italian males, where we obtained a negative correlation explaining about 20% of the variance which suggested that males with a higher education are characterized by a lower sense of community. However, the Italian females reported a positive correlation between the two measures explaining the 8% of the variance. For what concerns the privacy construct, education appears to affect only the Italian female perception. In particular, we always obtained a negative correlation accounting for 9% and 16% of the variance respectively for privacy confidence and privacy concerns.

The sense of community correlation structure shows interesting features as well. First of all no significant correlation appears to relate the sense of community with privacy and participation for the females disregarding nationality. On the contrary, the male sub-samples both reports positive correlations between sense of community and participation confidence with higher values for Italians ($r = 0.69$), and lower values for Turkish students ($r = 0.39$). While an opposite behavior appears to characterize the relation with the participation concerns, with values respectively of $r = 0.54$ for Italians, and of $r = -0.33$ for Turkish males. Only the Italian males show a significant positive correlation between sense of community and privacy concerns ($r = 0.58$). In order to evaluate the non-trivial relationships existing between concerns and confidence for both the participation and privacy perceptions, we explicitly took into account all the correlation structures describing such sub-measures separately.

The participation confidence appears to be significantly and negatively related to the participation concerns only within the Italian females sub-sample, explaining 28% of the variance. For what concerns the relationship between participation and privacy confidence, all the sub-samples report positive correlations with the only exception of the Italian males, suggesting how an increment in the first produces a coherent change in the second. In particular, the Turkish females show the higher correlation ($r = 0.61$), and the Turkish males the second one ($r = 0.40$), with the Italian females that report the weakest relation ($r = 0.28$). Finally the participation confidence is even related with the privacy concerns for Italian males ($r = 0.40$), and Turkish females ($r = 0.35$). In brief, the participation concerns are reported to correlate with the privacy concerns, and not with the privacy confidence for all the investigated sub-samples. All the correlations here are between 11% and the 22%.

2.2.2. Sense of Community: Nationality and Gender Effects

The effects of nationality and gender, as well as their interaction on the sense of community were evaluated by means of ANOVA analysis. As reported in Table 6, no differences emerge regarding the gender, nor the interaction between nationality and gender. The only factor affecting such a dimension appears to be the nationality, even if the effect is very moderate explaining just 3.5% of the variance, with the Italian sub-sample reporting a higher value.

Table 6. In table the best ANOVA model for the dependent variable Sense of Community, considering the factors gender and nationality, is reported.

Sense of Community			
Factor	F	*Sign*	η^2
Nationality	6.4	$p < 0.05$	0.035 (3.5 %)
Gender	1.2	ns	0.007 (0.7%)
Nat*Gen	1.5	ns	0.009 (0.9%)

2.2.3. Participation Dimensions: Nationality and Gender Effects

Confidence and concerns sub-measures concerning participation were separately analyzed by ANOVA (Table 7), in order to estimate the role of nationality and gender on the two dimensions. For both categories, gender and interaction vs. gender and nationality do not have statistically significant effects, while again nationality appears to be the only factor impacting on such dimensions. In both the cases, the cultural differences seem to account for approximately 10% of the variance. In particular, both the participation confidence as well as the participation concerns appear to be significantly higher within the Turkish sample.

Table 7. In table the best ANOVA model for the variables connected to the participation dimension (i.e., confidence & concerns) are presented, considering the factors gender and nationality.

Participation Confidence			
Factor	F	*Sign*	η^2
Nationality	20.7	$p < 0.01$	0.105 (10.5 %)
Gender	0.1	ns	0.001 (0.1%)
Nat*Gen	1.8	ns	0.010 (1%)
Participation Concerns			
Factor	F		η^2
Nationality	20.1	$p < 0.01$	0.102 (10.2 %)
Gender	0.1	ns	0.001 (0.1%)
Nat*Gen	0.6	ns	0.009 (0.9%)

2.2.4. Privacy Dimension: Nationality and Gender Effects

In Table 8 the best predicting model linking the nationality and gender factors to the sub-scales related to the privacy dimension (i.e., confidence and concerns) is reported. In these cases, we obtained a significant effect both on nationality as well as on gender, and again no effects were revealed about the interaction between the factors of interest. In particular, nationality explains 11.4% for the privacy confidence, and the 14.1% of the privacy concerns. In both cases, again the Turkish sample is characterized by higher values. In regards to gender, females reported higher values for privacy concerns, and lower values for privacy confidence, respectively explaining 8.3% and 2.1% of the total variance.

Table 8. In table the best ANOVA model for the variables connected to the privacy dimension (i.e., confidence & concerns) are presented, considering the factors gender and nationality.

Privacy Confidence			
Factor	F	*Sign*	η^2
Nationality	22.6	$p < 0.01$	0.114 (11.4 %)
Gender	15.8	$p < 0.05$	0.083 (8.3%)
Nat*Gen	1.3	ns	0.007 (0.7%)
Privacy Concerns			
Factor	F		η^2
Nationality	28.9	$p < 0.01$	0.141 (14.1 %)
Gender	3.7	$p < 0.05$	0.021 (2.1%)
Nat*Gen	0.1	ns	0.001 (0.1%)

2.2.5. Semantic Differential about Participation and Privacy

Very interesting results emerged from the semantic differential analysis regarding the nationality and gender as well as their combined effect. Table 9 reports only the significant effects, that nevertheless appear to be present in both cases (i.e., semantic differential related to privacy and participation separately) for 8 dimensions out of the original 10. Some dimensions are characterized by moderate effects, with an explained variance ranging between 2.6% and 9.6%. It is worth noting some dimensions appear as strongly affected by the nationality factor. For instance the concept of participation appears to elicit impressive differences, with "bad-good" accounting for 61.2% of the total variance, "dirty-clean" for 50.2%, and "useless-useful" for 46.5%. Even the privacy concept appears differently represented by the two cultural samples, with "personal-social" dimension explaining 45.8%, "useless-useful" 45.2%, "dirty-clean" 44.1%, and "bad-good" 40.4%. In particular, as reported in Figures 1 and 2, the Italian sample seems to perceive the participation concept as more "clean", "good" and "useful" compared to the Turkish sample. The same effects are reported for the privacy concept with the only exception of the dimension "personal-social", in which the Turkish sample perceives "privacy" as more social than the Italians.

Table 9. In table the best ANOVA models for the variables connected to the semantic differentials about both Participation and Privacy are presented, considering the factors gender and nationality. Statistical significance (*p*) is reported in table as follows: *: <0.05 and **: <0.01.

Participation Dimension	Factor	F	η^2
Warm-Cold	Nationality	4.7 *	0.026 (2.6 %)
	Nat*Gen	5.5 *	0.031 (3.1%)
Dirty-Clean	Nationality	177.6 **	0.502 (50.2 %)
Safe-Dangerous	Nat*Gen	7.3 **	0.040 (4.0%)
Bad-Good	Nationality	277.8 **	0.612 (61.2 %)
	Gender	12.3 **	0.065 (6.5%)
Useless-Useful	Nationality	153.0 **	0.465 (46.5 %)
	Gender	15.2 *	0.080 (8.0%)
Imp.-Unimportant	Nationality	7.4 **	0.041 (4.1 %)
	Nat*Gen	11.2 **	0.060 (6.0%)
Personal-Social	Nationality	18.8 **	0.096 (9.6 %)
	Gender	8.4 **	0.045 (4.5%)
Concrete-Abstract	Nationality	63.9 **	0.266 (26.6 %)
Privacy Dimension	**Factor**	**F**	η^2
Warm-Cold	Nationality	12.6 **	0.067 (6.7 %)
	Gender	4.5 *	0.025 (2.5%)
Dirty-Clean	Nationality	138.6 **	0.441 (44.1 %)
Bad-Good	Nationality	199.4 **	0.404 (40.4 %)
Fun-Boring	Nationality	9.1 **	0.049 (4.9 %)
Useless-Useful	Nationality	145.2 **	0.452 (45.2 %)
	Gender	4.2 *	0.023 (2.3%)
	Nat*Gen	5.2 *	0.028 (2.8%)
Imp.-Unimportant	Gender	4.1 *	0.023 (2.3 %)
Personal-Social	Nationality	148.6 **	0.458 (45.8 %)
	Nat*Gen	4.9 *	0.027 (2.7%)
Concrete-Abstract	Nationality	5.5 *	0.030 (3.0 %)

3. Discussion

The psychological constructs of Participation and Privacy are currently among the most interesting psychological concepts to investigate. The growing multi-disciplinary literature about them, ranging from engineering to complex systems and social sciences, witnesses the impact of ICT on modern human society, and in general on every new kind of interactions among people and groups [43–45]. Many societal dynamics, from the microscopic interaction within families to the macroscopic cultural and sociological movements involving people from all over the world, are based on the same human tendencies of interaction with others by means of the adoption of cultural and personal dependent strategies of coping. From the "Arab spring" in 2010, to the "Occupy Wall Street movements" in 2011, and the recent protests in Turkey (i.e., from Taksim square protests to the recent coup), the Internet has been considered by many commentators as one of the fundamental ingredients. Of course the connection between people, especially within the younger generations, has provided new incentives and mainstreams that have nurtured the protest movements. On the other hand, such events could be considered as a direct consequence of the changes produced by Internet on its users. Actually, from this point of view it seems very hard to clarify the role of the world wide web on such cultural and societal changes, while it seems more reasonable to take into account both the dynamics described above. Human beings change in order to satisfy environmental requirements (i.e., increasing their adaptation), to negotiate social constructs, norms and behaviors useful to communicate, interact and solve problems with others as well as represent elements affecting and actively changing the environment they belong to. For instance, the spreading of social networks in the last few years required non-digital natives to make a great effort to adapt to such a brand new social order, while it profoundly shaped the social skills and attitudes of digital natives. Such a double process (i.e., to adapt old strategies to new tasks, and to adopt new concepts and habits) is not just a question of age (i.e., digital natives vs non-digital natives), but it represents a double process acting in different degrees within every participant, and in different ways (i.e., with different speed and effectiveness) along the different cultures, and socio-demographic clusters. In general, it is expected that the dynamics of such changes will define new interesting scenarios and phenomena, and of course both the spreading of new concepts and social negotiations of old habits and norms will continue to determine societal as well as personal changes.

The hypotheses of our study moved from the basic assumption that the "fluid" human perception of socially-negotiated concepts like Participation and Privacy, can differ a lot between two different cultural groups, depending on structural differences of the cultures and on different recent events affecting the university students participating in the survey.

The Turkish and the Italian cultures have ancient common roots, and could be both considered as European cultures even if with a very different recent history. Finally, the different role of gender within the two cultural groups was considered and evaluated in the multivariate analysis.

The two sub-samples appeared as comparable regarding the size, average age and gender balance. With the age that actually presented very little dispersion around the average values, and that can be considered as a constant describing a very specific social cluster (i.e., those of the university students from a medium size city). Consequently, it is not surprising that no significant correlations emerged concerning age. The only correlation with the age variable is reported in the Turkish female sub-sample as affecting the Privacy Confidence. The older the participant is the greater is her confidence and trust in the concept of Privacy (Table 5).

The education variable represents the number of formal years of study accomplished by the participants, and the different distribution of the two sub-samples is due only to the different organization of the two educational systems. For this reason only the Italian sample presents significant correlations between the education variable and the operative dimensions. The opposite relationship between gender and sense of community is very interesting. In the Italian male sub-sample a negative correlation is revealed, while in the female one a significant positive correlation is present. Moreover in the Italian sample, the males are characterized by a negative correlation between sense of community and education, while the Italian females show the opposite trend. The Privacy sub-dimensions, confidence and concerns, show a negative correlation in age only for Italian females, and consequently appear to reduce both their confidence and concerns about Privacy with the increase of education.

The sense of community dimension shows very interesting relation patterns within our sample, reporting significant correlations only for the male sub-samples, and not always in the same direction. The Participation concept is perceived as more positive and trustworthy as the sense of community increases, the same effect acts on Italian male sub-samples even for the concerns, while it presents an opposite effect on the Turkish male sub-sample (i.e., the concerns about participation decrease with the increase of sense of community). Finally the Italian sub-sample presents a positive correlation between the sense of community and Privacy concerns scores (Table 5).

In general, it is interesting to observe how the results of the correlation matrix between Participation and Privacy measurements are quite empty. In particular, larger correlations were expected between the confidence and concerns sub-measures. Such data indicates the multi-structured and complex architecture of the concepts under scrutiny and how in different cultures their interplay could not be easy to reveal or predict.

In order to evaluate the interplay between gender and culture on the study variables of our study (i.e., sense of community, participation and privacy concerns and confidence), we conducted a series of separate ANOVA in order to test the effects of the factors as well as of their interaction.

The sense of community does not appear to differ much regarding the nationality (3.5%), and the gender does not play a significant role in the interaction between gender and nationality. In other words, the two cultures appear to be comparable in terms of sense of community.

On the contrary, more relevant differences are detected regarding the Participation and Privacy measures. The ANOVA investigating the gender and nationality effects on the sub-scores of the Participation scale (Table 7) reports that around the 10% of the variance is explained by the nationality factor. In particular, both the confidence and the concerns about participation are significantly higher in the Turkish sample. A greater portion of the variance is explained by the ANOVA investigating such effects on the Privacy sub-scores. Concerning this construct respectively 11.4% and the 14.1% are associated with the nationality, again indicating the Turkish sub-sample as more confident and concerned about Privacy (Table 8). Finally an effect of the gender is revealed as explaining respectively the 8.3% and the 2.1% of the variance of Privacy confidence and concerns sub-scores, with the Females reporting always a lower score.

The most fascinating results are those produced by the semantic differential measures. Even in this case a series of ANOVA were conducted analyzing every dimension presented by the semantic differential, considering the gender and the nationality as independent factors of variance. The Fisher's Fs and the Etas squared reported in table 8 delineates impressive differences in the perception of such concepts always mainly related to the nationality (i.e., the culture).

Both the perception of the Participation and Privacy appear to be affected by nationality and gender; in particular, in 8 cases out of 16 the explained variance (i.e., the entity of the difference explained by the factors) is under 10%. Surprisingly the other 8 cases report an explained variance ranging between the 61.2% and the 26%.

For what concerns Participation, it seems that Italians perceive it as a more "clean", "good", "useful" and "social" than Turkish, with no significant difference to gender. Among the samples, there are no significant differences on the perceived importance, warmness, easiness and perceived "fun" related to the experience of participating. The fact that Turkish people perceive it as more "dirty", "concrete", "bad", "useless" could be related to the recent experiences of the Taksim square protests in which really violent clashes happened.

As regards to the concept of Privacy, it seems that the Italians perceive it as a more "useful", "good", "clean" and "personal" concept than the Turkish students. In order to appreciate these results with a cultural interpretation, we wanted to highlight the key role that the recent Italian mass media and political debates have given to the concept of privacy. In fact, the Italian political agenda in the recent years has been highly focused on a privacy regulation, especially regarding important legal topics. This concern has led to a more developed concept of privacy, highly related to the concept of "human rights that need to be respected and protected" as found in the definition provided by the main Italian dictionary (Treccani dictionary). This concept of privacy as a right is not stressed at the same level for example in the current English definition, which stresses more the perspective of an "ability of an individual or group to seclude themselves" (Wikipedia), demonstrating how this concept is still in development and culturally shaped and variable. The fact that the Turkish students perceive the construct of Privacy as more "useless", "social", "dirty" and "bad" could be given to the fact that this concept is quite new to them. In fact, the word "privacy" is not in the main Turkish dictionaries at the moment and even the literal translation of this term in this language ("gizlilik") needed a well-structured explanation to make it comprehensible to the Turkish sub-sample. We believe that a better understanding of the dynamics underneath the use of ICT tools can provide us the necessary knowledge in order to shape these environments in a more efficient way (i.e., taking into account cultural, gender and historical differences), strengthening positive features and mitigating the negative ones.

Main limitation to this study is the fact that the questionnaires aimed to assess Participation and Privacy constructs are used for the first time. Nevertheless, since measures for both constructs are actually lacking, authors believe that those instruments should be taken in consideration for next studies. Another important limitation for the study is the lack of an instrument assessing cultural traits, especially to show differences and similarities between Mediterranean cultures. Authors decided to refer directly to nationality since the recent political escalations in Turkey and Italy provided interesting contexts for the research. Another relevant limitation to the used instruments is that studies using Sense of Community measures suggest caution in the interpretation because of the nature of the construct.

4. Conclusions and Future Work

While ICT tools are becoming everyday more important in our lives, a proper analysis of their effects on societal dynamics is still developing. The increasing use of social media as a tool to communicate with large groups and the possibility for everyone to use them to broadcast messages and opinions all over the world are changing our ways of thinking about information, providing brand new ways to interpret reality and previous limitations to shared knowledge. Social media are in fact playing a crucial role in shaping not only our daily routine but also worldwide assets. While this paper was still in development the political asset in so many different countries has changed so quickly that we couldn't imagine just 10 years ago: from the "Arab Spring" movements to the more recent anti-coup movements in Turkey ICT has provided the tools for many people to stand up and make their voices heard. We wanted to assess whether the use of this tools and the role social media had during recent historical events shaped the perception that people have about them. In particular, we developed two scales to assess the perception of online participation and online privacy in order to estimate in which case people are perceiving ICT use as a threat or as a resource while engaging debates online. To understand this historic change in communication, it is important for us to estimate not only the use of internet tools but also how the use of ICT tools is perceived by users. It is in fact reasonable to think that the recent history of a country, in which social media had a key role in shaping the national asset, shaped the perception that people living in that country have of these tools in terms of confidence and concerns about their use. Standing to this, users' perception on ICT's trustworthiness could somehow provide us an indicator of the level of democracy and freedom experienced by people, or even of the probability that these people have to organize social movements online. This paper represents, of course, just a first step in this direction, but future perspectives and developments are fascinating and could benefit of the raising literature on the subject. Future works could, of course, take into account different cultures from the two examined in this paper. Another interesting development could be made by connecting the social dimensions assessed in this paper with the cultural dimensions provided by Hofstede (CITARE) in order to link features assigned to ethnicity to a more precise cultural dimension.

Acknowledgments: This work was funded by the European Commission under the FP7-ICT-2013-10 call, proposal No. 611299, project SciCafe 2.0, and by H2020 FETPROACT-GSS CIMPLEX Grant No. 641191. D.V. acknowledges the support from the project CLARA (CLoud plAtform and smart underground imaging for natural Risk Assessment), funded by the Italian Ministry of Education and Research (PON 2007-2013: Smart Cities and Communities and Social Innovation; Asse e Obiettivo: Asse II - Azione Integrata per la Società dell'Informazione; Ambito: Sicurezza del territorio), and from H2020 FETPROACT-GSS CIMPLEX, Grant No. 641191.

Author Contributions: A.G. conceived the study and coordinated the group research; C.D. and A.N. collected the Italian data; A.S. collected the Turkish data; P.M. supervised the study; A.G. and D.V. accomplished the data analysis; A.G., C.D., P.M. and D.V. wrote the paper; D.V. wrote the revisions.

Conflicts of Interest: The authors declare no conflict of interest.

Appendix A

Participation Scale

General Instructions: Dear participant as follows you will find a list of ten dimensions, each of them bound by two opposite adjectives potentially describing the concept of "Participation". Please provide a check on a single value along the dimension (from +5 on left to +5 on right) for each dimension, filling in the entire chart. In order to answer properly, please consider the vocabulary definition of the word "Participation" reported below. Thank you.

Participation Definition: Participation refers to different mechanisms for the public to express opinions—and ideally exert influence—regarding political, economic, management or other social decisions. Participation can take place along any realm of human social activity, including economic (i.e., participatory economics), political (i.e., participatory democracy or parpolity), management (i.e., participatory management), cultural (i.e., multiculturalism) or familial (i.e., feminism).

Please answer the following questions regarding your opinion about Participation. Please refer to your personal knowledge and try to describe your general approach instead of your specific past experiences. If you have no previous experience regarding participation, please refer to the given definition.

(Answer style: 1: absolutely no, 2: a little, 3: moderately, 4: a lot, 5: very much)

- 1. I think that engaging personally in something is in general costly and frequently not efficient.
- 2. I like being part of my community life.
- 3. Whenever I can, I prefer to delegate important things to others.
- 4. Everyone should be actively engaged in the administration of their world.
- 5. To be participative on the web is not frequently effective.
- 6. The web represents a great opportunity to activate people concerning important topics.
- 7. Social networks engagement is risky because some people could unfairly exploit the efforts of others.
- 8. Initiatives promoted within social networks could make the world a better place.
- 9. Too much active participation on social networks can be psychologically and socially damaging.
- 10. Participating in social networks makes me connect with people who improve the quality of my life.

Scoring: Even questions are about Concerns, while odd questions are about affordance. Consequently the scores of the two subscales are obtained just summing the two series of items as follows:

Participation Concerns: 1 + 3 + 5 + 7 + 9
Participation Affordance: 2 + 4 + 6 + 8 + 10

Privacy Scale

General Instructions: Dear participant as follows you will find a list of ten dimensions, each of them bound by two opposite adjectives potentially describing the concept of "Privacy". Please provide a check on a single value along the dimension (from +5 on left to +5 on right) for each dimension, filling in the entire chart. In order to answer properly, please consider the vocabulary definition of the word "Privacy" reported below. Thank you.

Privacy Definition: Privacy is the ability of an individual or group to seclude themselves, or information about themselves, and thereby express themselves selectively. The boundaries and content of what is considered private differ among cultures and individuals, but share common themes. When something is private to a person, it usually means that something is inherently special or sensitive to them. The domain of privacy partially overlaps security, which can include the concepts of appropriate use, as well as protection of information. Sometimes, defined as the most private part of a person's life, that he or she has the right to defend.

Please answer the following questions regarding your opinion about Privacy. Please refer to your personal knowledge and try to describe your general approach instead of your specific past experiences. If you have no previous experience regarding privacy, please refer to the given definition.

(Answer style: 1: absolutely no, 2: a little, 3: moderately, 4: a lot, 5: very much)

- 1. I am concerned about sharing my personal information.
- 2. I usually disclose certain information about myself easily.
- 3. I am worried about someone using my personal data.
- 4. Providing information about myself is useful for my social life.
- 5. I think it is very dangerous to share personal codes on the web (i.e., credit card number, user passwords)
- 6. I am able to manage properly my information on the web.
- 7. On social networks I am only connected to people I can trust.
- 8. Anyone can view what I share on my social networks.
- 9. Information on my social networks can be recovered and used to damage me.
- 10. It is more comfortable to share personal information on social networks than in real life.

Scoring: Even questions are about Concerns, while odd questions are about affordance. Consequently the scores of the two sub-scales are obtained just summing the two series of items as follows:

Privacy Concerns: 1 + 3 + 5 + 7 + 9
Privacy Affordance: 2 + 4 + 6 + 8 + 10

References

1. Dunbar, R.I. Neocortex size as a constraint on group size in primates. *J. Hum. Evol.* **1992**, *22*, 469–493.
2. Dunbar, R.I. The social brain hypothesis and its implications for social evolution. *Ann. Hum. Biol.* **2009**, *36*, 562–572.
3. Bronfenbrenner, U. *Ecological Systems Theory*; Jessica Kingsley Publishers: London, UK, 1992.
4. Prensky, M. Digital natives, digital immigrants part 1. *Horizon* **2001**, *9*, 1–6.
5. Arnaboldi, V.; Guazzini, A.; Passarella, A. Egocentric online social networks: Analysis of key features and prediction of tie strength in Facebook. *Comput. Commun.* **2013**, *36*, 1130–1144.
6. Passarella, A.; Dunbar, R.I.; Conti, M.; Pezzoni, F. Ego network models for future internet social networking environments. *Comput. Commun.* **2012**, *35*, 2201–2217.
7. Brodsky, A.E.; Loomis, C.; Marx, C.M. Expanding the conceptualization of PSOC. In *Psychological Sense of Community*; Springer: Berlin, Germany, 2002; pp. 319–336.
8. Brodsky, A.E.; Marx, C.M. Layers of Identity: Multiple Psychological Senses of Community within a Community Setting. *Psykhe* **2011**, *13*, doi:10.4067/S0718-222820040002000184.
9. Sarason, S.B. *The Psychological Sense of Community: Prospects for a Community Psychology*; Jossey-Bass: Hoboken, NJ, USA, 1974.
10. Blanchard, A.L. Testing a model of sense of virtual community. *Comput. Hum. Behav.* **2008**, *24*, 2107–2123.
11. McMillan, D.W.; Chavis, D.M. Sense of community: A definition and theory. *J. Community Psychol.* **1986**, *14*, 6–23.
12. Castellini, F.; Colombo, M.; Maffeis, D.; Montali, L. Sense of community and interethnic relations: Comparing local communities varying in ethnic heterogeneity. *J. Community Psychol.* **2011**, *39*, 663–677.
13. Blanchard, A.L.; Markus, M.L. Sense of virtual community-maintaining the experience of belonging. In Proceedings of the 35th Annual Hawaii International Conference on System Sciences, Washington, DC, USA, 7–10 January 2002; pp. 3566–3575.
14. Rovai, A.P.; Wighting, M.J. Feelings of alienation and community among higher education students in a virtual classroom. *Internet High. Educ.* **2005**, *8*, 97–110.
15. Francescato, D.; Porcelli, R.; Mebane, M.; Cuddetta, M.; Klobas, J.; Renzi, P. Evaluation of the efficacy of collaborative learning in face-to-face and computer-supported university contexts. *Comput. Hum. Behav.* **2006**, *22*, 163–176.
16. Francescato, D.; Mebane, M.; Porcelli, R.; Attanasio, C.; Pulino, M. Developing professional skills and social capital through computer supported collaborative learning in university contexts. *Int. J. Hum.-Comput. Stud.* **2007**, *65*, 140–152.

17. Cicognani, E.; Albanesi, C.; Zani, B.; Mazzoni, D.; Bertocchi, A.; Villano, P. Civic and political participation across generations in Italy: A qualitative study. In Proceedings of the Bologna PIDOP Conference 2011, University of Bologna, Bologna, Italy, 11–12 May 2011.

18. Wilkenfeld, B.; Lauckhardt, J.; Torney-Purta, J. The relation between developmental theory and measures of civic engagement in research on adolescents. In *Handbook of Research on Civic Engagement in Youth*; Wiley: Hoboken, NJ, USA, 2010; pp. 193–220.

19. Klandermans, B.; Staggenborg, S. *Methods of Social Movement Research*; U. of Minnesota Press: Minneapolis, MN, USA, 2002; Volume 16.

20. Simon, B.; Loewy, M.; Stürmer, S.; Weber, U.; Freytag, P.; Habig, C.; Kampmeier, C.; Spahlinger, P. Collective identification and social movement participation. *J. Personal. Soc. Psychol.* **1998**, *74*, 646.

21. Rappaport, J.; Simkins, R. Healing and empowering through community narrative. *Prev. Hum. Serv.* **1991**, *10*, 29–50.

22. Zimmerman, M.A. Empowerment theory. In *Handbook of Community Psychology*; Springer: New York, NY, USA, 2000; pp. 43–63.

23. Gächter, S.; Renner, E.; Sefton, M. The long-run benefits of punishment. *Science* **2008**, *322*, 1510–1510.

24. Rand, D.G.; Dreber, A.; Ellingsen, T.; Fudenberg, D.; Nowak, M.A. Positive interactions promote public cooperation. *Science* **2009**, *325*, 1272–1275.

25. Sutter, M.; Haigner, S.; Kocher, M.G. Choosing the carrot or the stick? Endogenous institutional choice in social dilemma situations. *Rev. Econ. Stud.* **2010**, *77*, 1540–1566.

26. Gächter, S. Social science: Carrot or stick? *Nature* **2012**, *483*, 39–40.

27. Tsiatsikas, Z.; Geneiatakis, D.; Kambourakis, G.; Keromytis, A.D. An efficient and easily deployable method for dealing with DoS in SIP services. *Comput. Commun.* **2015**, *57*, 50–63.

28. Tene, O. Privacy: The new generations. *Int. Data Priv. Law* **2011**, *1*, 15–27.

29. Mivule, K. Web Search Query Privacy, an End-User Perspective *J. Inf. Secur.* **2017**, *8*, 1.

30. Wang, C.; Shi, D.; Xu, X. AIB-OR: Improving Onion Routing Circuit Construction Using Anonymous Identity-Based Cryptosystems. *PLoS ONE* **2015**, *10*, e0121226.

31. Pfitzmann, A.; Hansen, M. A Terminology for Talking About Privacy by Data Minimization: Anonymity, Unlinkability, Undetectability, Unobservability, Pseudonymity, And Identity Management, 2010. Available online: http://www.maroki.de/pub/dphistory/2010_Anon_Terminology_v0.34.pdf (accessed on 7 April 2017).

32. Song, Z.; Hao, C.; Daqing, Z. Empirical study on users' participation behavior in SNS based on theory of perceived risks and involvement degree. In Proceedings of the 2013 10th International Conference on Service Systems and Service Management (ICSSSM), IEEE, Hong Kong, China, 17–19 July 2013; pp. 424–429.

33. Solove, D.J. The end of privacy? *Sci. Am.* **2008**, *299*, 100–106.

34. Altman, I. *The Environment and Social Behavior: Privacy, Personal Space, Territory, and Crowding*; Brooks/Cole Pub. Co.: Pacific Grove, CA, USA, 1975.

35. Pedersen, D.M. Psychological functions of privacy. *J. Environ. Psychol.* **1997**, *17*, 147–156.

36. Rustemli, A.; Kokdemir, D. Privacy dimensions and preferences among Turkish students. *J. Soc. Psychol.* **1993**, *133*, 807–814.

37. Kaya, N.; Weber, M.J. Cross-cultural differences in the perception of crowding and privacy regulation: American and Turkish students. *J. Environ. Psychol.* **2003**, *23*, 301–309.

38. Arpaci, I.; Kilicer, K.; Bardakci, S. Effects of security and privacy concerns on educational use of cloud services. *Comput. Hum. Behav.* **2015**, *45*, 93–98.

39. Osgood, C.E. Semantic differential technique in the comparative study of Cultures1. *Am. Anthropol.* **1964**, *66*, 171–200.

40. Chow, S.C.; Wang, H.; Shao, J. *Sample Size Calculations in Clinical Research*; CRC Press: Boca Raton, FL, USA, 2007.

41. Rovai, A.P.; Wighting, M.J.; Lucking, R. The classroom and school community inventory: Development, refinement, and validation of a self-report measure for educational research. *Internet High. Educ.* **2004**, *7*, 263–280.

42. Vieno, A.; Santinello, M.; Pastore, M.; Perkins, D.D. Social support, sense of community in school, and self-efficacy as resources during early adolescence: An integrative model. *Am. J. Community Psychol.* **2007**, *39*, 177–190.

43. Castellano, C.; Fortunato, S.; Loreto, V. Statistical physics of social dynamics. *Rev. Mod. Phys.* **2009**, *81*, 591.
44. Vilone, D.; Carletti, T.; Bagnoli, F.; Guazzini, A. The Peace Mediator effect: Heterogeneous agents can foster consensus in continuous opinion models. *Phys. A Stat. Mech. Its Appl.* **2016**, *462*, 84–91.
45. Cecconi, F. *New Frontiers in the Study of Social Phenomena*; Springer: Heidelberg, Germany, 2016.

MDPI AG

St. Alban-Anlage 66

4052 Basel, Switzerland

Tel. +41 61 683 77 34

Fax +41 61 302 89 18

http://www.mdpi.com

Future Internet Editorial Office

E-mail: futureinternet@mdpi.com

http://www.mdpi.com/journal/futureinternet

www.ingramcontent.com/pod-product-compliance
Lightning Source LLC
Chambersburg PA
CBHW041216220326
41597CB00033BA/5982